Vadim Gordienko
Ivan Gordienko
Olga Zavgorodnyaya

Thermal field and geoenergy resources of Ukraine

Vadim Gordienko
Ivan Gordienko
Olga Zavgorodnyaya

Thermal field and geoenergy resources of Ukraine

ScienciaScripts

Imprint

Cover image: www.ingimage.com

This book is a translation from the original published under ISBN 978-3-659-85303-6.

Publisher:
Sciencia Scripts
is a trademark of
Dodo Books Indian Ocean Ltd. and OmniScriptum S.R.L publishing group

120 High Road, East Finchley, London, N2 9ED, United Kingdom
Str. Armeneasca 28/1, office 1, Chisinau MD-2012, Republic of Moldova, Europe
Managing Directors: Ieva Konstantinova, Victoria Ursu
info@omniscriptum.com

Printed at: see last page
ISBN: 978-620-8-37369-6

Table of Contents

Introduction

The thermal field of the continents and oceans has received much attention in recent decades, and geothermal data are increasingly used in analyzing fundamental problems of Earth sciences and solving applied problems. They occupy a unique position among the data of other geophysical methods, as they allow us to directly study processes in the subsurface, rather than their more or less remote consequences. However, the intensity of experimental and theoretical studies of the thermal field does not correspond at all to such a role. The density of the network of geothermal gradient and heat flux (HF) determinations lags far behind that achieved in the study of other physical fields. There are no generally accepted requirements for the processing of observed TP and field analysis, which would ensure the adequacy of the resultant models to the real conditions in the subsurface and physical laws. Therefore, despite significant computerization of data processing and interpretation, mutually exclusive models based on the same initial information are often encountered. Speculative and qualitative interpretations of the thermal field are still widespread, with the help of which it is possible to "harmonize" TP anomalies with any deep processes and objects.

In studying the thermal field of Ukraine, two tasks were set and largely solved.

1 The detail and reliability of the heat flux study were brought to a level that allows the isolation of the main disturbances on a regional scale.

2 . The system of experimental data analysis, which is maximally close to the unambiguous solution of the inverse problem, i.e., the construction of geologically and geophysically reliable thermal models of the crust and upper mantle, was created.

To achieve these goals, a network of heat flux determinations (mainly - in the last 10-20 years) was created, including TP values in 13000 wells (one well has from 1 to 20 values). All of them were corrected taking into account the influence of paleoclimate, non-horizontal groundwater overflows, geological structures (when the latter led to non-horizontal boundaries of interfaces between media with different thermal conductivity), young thrusts, intensive young sedimentation The error of the determined values is estimated at an average of 5-10%. The introduced corrections made it possible to completely exclude changes in the heat flux by depth in one well, noticeably exceeding the calculation error. As a result, the Map of the deep heat flux of Ukraine was constructed. In terms of network density it has no analogues among the maps for comparable areas.

The use of a large amount of information on the crustal structure and composition of its rocks, as well as their radiogenic heat generation, made it possible to fully explain the background values of the heat flux in inactivated platform regions. At the same time, the TP value from the mantle is justified by the results of independent consideration of its thermal history.

The difficulties in interpreting the thermal field (ambiguity of the inverse problem solution, "reaching a maximum", for example, in areas where the effect of a young heat source did not propagate to the surface and did not manifest itself in the TP) are seemingly insurmountable. However, it is the dependence of the observed effect on a multitude of factors that prompts us to think about the possibility of building a strong scheme for regularizing the solution of the inverse problem and externally controlling the results of such a solution.

Chapter 1: Determination of the Earth's heat flux on the territory of Ukraine

The main parameter studied in modern geothermy is the Earth's heat flux. Calculation of the observed value of TP requires knowledge of the geothermal gradient (γ) and thermal conductivity (λ) of rocks in one depth interval (TP = $\gamma\lambda$). Therefore, it is necessary to dwell on the characteristics of these quantities, considering the option of research in relatively deep boreholes (the method of temperature wave reduction - RTW, which does not require deep boreholes and which establishes a small part of TP, is described in [17]).

1.1 Determination of temperature in wells

When evaluating the applicability of T measurements in the borehole to calculate the geothermal gradient with acceptable accuracy, it is necessary to know the measurement error and the correspondence of the measured temperature to the natural temperature.

Special work was done to find out the error level of T determinations based on the results of repeated measurements in wells [14]. Only data on mature wells were used. Their depths ranged from 200 to 2500 m. The boreholes penetrated mainly sedimentary rocks, and crystalline rocks in smaller quantities. Geothermal gradients varied in wide ranges - 1-40^0 C/100m. Depth intervals with obvious distortions of near-surface nature were excluded from consideration. The temperatures established by the specialists of the Institute of Geophysics of the Academy of Sciences of Ukraine, the Institute of Marine Geology and Geophysics FEB RAS, the Institute of Seismology of the Academy of Sciences of Kyrgyzstan, the Institute of Physics of the Earth RAS, the Institute of Geochemistry and Geophysics of the Academy of Sciences of Belarus, Kazan University, the Institute of Mineral Resources of the Ministry of Geology of Kazakhstan, VSEGEI, VIRG, St. Petersburg Mining Institute, as well as various GIS units of the Ministry of Geology of Russia were used.

The studied wells are located on the southern slope of the Baltic Shield, in the Pripyat Trough, the Dnieper-Donets Depression (DDD), on the slope of the Ukrainian Shield, in the Carpathians, the Transcarpathian Trough, the Crimea, the Precaucasus, the Turan Plate, the Tien Shan, the Kazakh Shield, Sakhalin and the Kuril Islands. A total of 90 wells were investigated and 250 thermograms were obtained (from two to nine in each well).

Temperature measurements made by precision thermometers were first considered. Several selections can be made from the available material.

1. Closely spaced measurements revealing insignificant temperature differences. The RMS difference is about 0.08^0 C. This value is close to that explained by the instrumental errors of the two thermometers, which demonstrates the possibility of achieving such a result in real well conditions.

2. Close in time (from several hours to a week) measurements, revealing relatively small differences in absolute temperatures that do not lead to significant errors in the geothermal gradient (no more than 5%). The RMS difference is about 0.3^0 C, which sharply exceeds the instrumental error of thermometers. Temperature changes in the wells during the period between measurements are unlikely.

3. Closely spaced measurements revealing drastic differences in results leading to significant errors in determining the magnitude of the geothermal gradient.

The RMS difference is about 1.70C. Differences γ reach 15% in large depth intervals (more than 1000 m) and 40-50% in small intervals (about 100 m).

In some cases, rapid changes in the performance of one of the thermometers used for comparison are evident. In wells in one of the exploration areas, temperatures measured by one thermistor thermometer differed from the logged temperatures by 5^0 C over several days. Subsequent monitoring with the same thermistor showed good agreement with the logs.

Thus, the real measurement error of an accurate thermometer may be large, but an error close to the instrumental error is also possible. The situation as a whole (for the period up to and including 1990, after which no such large-scale operation was carried out) is characterized by a summary histogram for all three samples. The standard deviation is about 0.7^0 C, but the distribution is far from normal, with frequent differences exceeding the triple standard deviation.

4. Repeated measurements made after a long time (from several months to 12 years) usually reveal noticeable differences in T. It should be noted that about 20% of repeated measurements were made with the same sensors used for measurements close in time. In the vast majority of cases, changes in T do not lead to significant changes in the geothermal gradient. The RMS value of the difference is the same 0.7^0 C. But, on the one hand, a significant part of the data included in sample 4 is established with the help of sensors with which samples 1 and 2 were obtained at close in time measurements. On the other hand - in some cases it was possible to assume that between the measurements there were changes in T of technogenic nature. Their magnitude could not be determined, but it can be significant, as follows from some cases studied in detail. For example, T changes were recorded at a depth of 800m in wells in the exploration area, where another well was drilled not far from them. The pumping carried out in it resulted in a sharp and prolonged change of T.

Excluding such data from sample 4, we obtain for it a histogram characterized by a RMS difference of 0.5°C. Thus, temporal changes in temperatures in the wells included in the analysis do not create significant anomalies of geological nature. In any case, they cannot be reliably distinguished at the achieved accuracy of T measurements.

Let us consider the reasons for the noted errors of temperature measurement. In some cases, they are related to unstable sensor characteristics, but such distortions are quite rare: only four thermograms out of 200 can be attributed to them. It is more likely that the widespread distribution of stable in time, but incorrectly determined characteristics of sensors is more probable. To test this assumption, sensors used by a number of scientific organizations conducting geothermal research in the 1990s were compared. All selected sensors were compared with the one used by the authors. The results obtained by them agree quite well with most of the established by other devices. Differences in absolute values of T in the first tenths of a degree were encountered, but they, as a rule, should not lead to significant differences in geothermal gradients. However, there were also sharp differences, reaching several degrees and clearly creating a danger of incorrect measurement of the geothermal gradient.

The use of such poor-quality sensors may well explain the established temperature discrepancies in the wells. The appearance of incorrect thermometer characteristics can logically be attributed to poor calibration. Analysis of numerous results of repeated calibrations performed by different geothermal groups allowed to find, along with the bulk of closely agreeing, and divergent calibrations. Their use provides at unchanged thermometer readings error T from a few tenths of a degree to several degrees and noticeable errors in determining the geothermal gradient.

Consider the available data on repeated temperature measurements using standard logging thermometers.

Both measurements close in time and those separated by significant periods of time are compared. Based on their results, a histogram of discrepancies is plotted, revealing a RMS value of this parameter of 1.2°C. It should be noted that in most cases subparallel thermograms were encountered, i.e., differences in absolute temperature values do not lead to differences in the values of geothermal gradients.

All considered pairs of thermograms were brought to the same level of absolute temperatures. The histogram of the preserved differences reveals a rms value of 0.3-0.4°C. Only in some depth intervals of several wells differences γ of about 25% were recorded. Typical values are about 10%. Some of the significant distortions of the gradient can presumably be attributed to the influence of anthropogenic factors. Temporal changes in temperatures in wells idle for a long time between measurements were not reliably detected in this group of results either.

Comparison of thermograms obtained by accurate thermometers with the thermo logging data was also carried out. The RMS value of the differences is 1.5°C. There are significant differences in γ in about half of the studied depth intervals. The larger discrepancy value compared to the one obtained when comparing the logging data with each other is explained by the fact that the results obtained by "accurate" thermometers with a large error, included in sample 3, were partially used in the sample under consideration. Thus, the quality of temperature measurements with standard logging thermometers may be higher than with "accurate" thermometers.

The conducted analysis allows us to draw the following conclusions.

1. Accurate thermometers used in geothermy in real well conditions allow measurements with an absolute temperature error of a few hundredths to the first tenth of a degree. They provide determination of the geothermal gradient with an error of 1-3%.

2. These hardware capabilities are often not utilized due to poor thermometer graduation. Therefore, significantly large errors are common: in absolute temperature - 0.5- l.0°C, in geothermal gradient - 10-15 %.

3. Determinations of the geothermal gradient from temperature measurements with standard logging thermometers do not differ significantly in accuracy from determinations from thermograms obtained by about half of the precision thermometers actually in use.

4. To improve the level of reliability of geothermal well surveys, periodic

reconciliation of the instruments used is necessary to identify and eliminate the results of incorrect graduations and other sources of large errors in temperature measurement.

The authors made their own measurements of temperature in standing wells using special thermistor thermometers. These measurements were made in about 5-10% of all wells for which T was used for TP calculation. We used thermistors manufactured in 1960 (i.e. naturally "aged") with practically no graduation change with time, with nominal resistance of 100000 Ohm. Control of characteristics (dependence of the sensor resistance on temperature) was carried out annually before the field season, mercury thermometers with an error of about $0.01\text{-}0.02^0$ C were used for calibration. The real error of thermometers was checked by repeated comparison of the results of consecutive measurements of T in one well. The average difference between them was $0.02\text{-}0.03^0$ C.

The same control results were obtained by the authors (VIRG staff) of T studies in shallow boreholes in the central part of the Ukrainian Shield (about 10-15% of all data used). Electric thermometers of their own design were used. It was also shown that the applied drilling technology (without circulation of flushing fluid) and short time of well sinking resulted in a very fast settling out. Already a few hours after drilling was completed, unchanged bottomhole temperatures were observed for several days.

Most of the heat flux determinations were based on bottomhole T measurements with standard logging thermometers during logging. As a rule, wells were stagnant from several hours to several days. The instrumental error of such an operation has been repeatedly estimated [14 and others] and amounts to the first tenths of a degree. The main error is clearly related to the distortions T introduced by the process of circulation of the flushing fluid during drilling. It is experimentally shown that the bottomhole temperature experiences the least distortion. It is actually measured several hours (and more) after the fluid circulation procedure is completed. Comparison with the data on aged wells convinces that in this period bottomhole T has already passed the period of exceeding the natural temperature associated with heat release during rock fracture. The degree of its proximity to the natural temperature is determined by drilling technology.

For sinking of the most part (about 55% of the total number) of the used wells (coal and hydrogeological) the technology was used, which does not lead to big distortions of bottomhole temperature. On examples of temperature measurements in 150 boreholes in Donbass it was shown that differences of T at bottom-hole temperatures from those obtained at the same depth and at a small distance (several hundred meters) during borehole measurements in fresh underground workings correspond to an error of about 0.3^0 C. Thus, the error did not exceed the instrumental error, and the natural T of the face had time to recover.

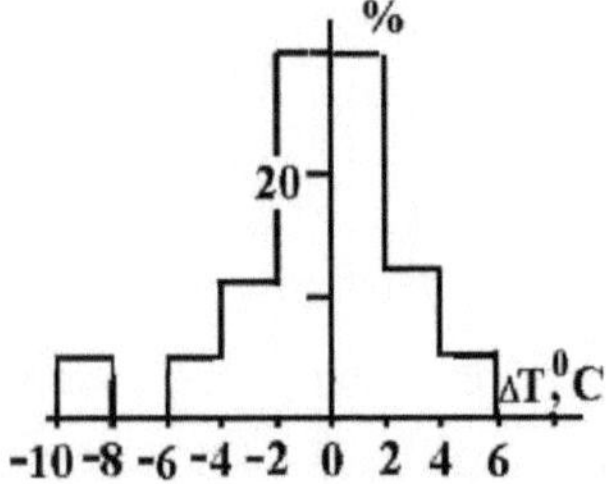

Fig. 1.1 Histogram of temperature differences in gas-bearing and water-bearing or oil-bearing formations.

Oil and gas sinking technology

wells (about 25% of all used wells) causes much more significant (about an order of magnitude) distortions of T, and the sign of the resulting anomaly and its dependence on bottomhole depth cannot be established.

Therefore, the results of such T measurements were used only for calculating TP in deep wells (more than 1000 m), which provided an acceptable (about 10% or less) error in calculating the average geothermal gradient between bottomhole and surface.

In recent years, the authors have developed a new methodology that allows to use for calculation of geothermal temperature gradient determined during tests of potentially productive formations in oil and gas wells. T measurements were mainly performed by the organizations of the Ministry of Geology of the Ukrainian SSR in the 70-80s of the twentieth century.

A methodological problem can be considered on these temperature data. There were fears that the opening of gas deposits during tests can lead to a noticeable temperature drop due to the throttle effect, creating an anomaly that distinguishes the value of the measured parameter from the natural one. In Donbass it is possible to compare the temperatures obtained during testing of gas-bearing formations with water-bearing and (attracting also the material on parts of the DDS) oil-bearing formations at the same depths within the same exploration areas.

The results of such comparison in the form of $\Delta T = T_{г} - T_{,вн}$ (where $T_{г}$ is the temperature in the gas-bearing reservoir, $T_{,вн}$ is the temperature in the water-bearing or oil-bearing reservoir) are shown in Fig. 1.1.

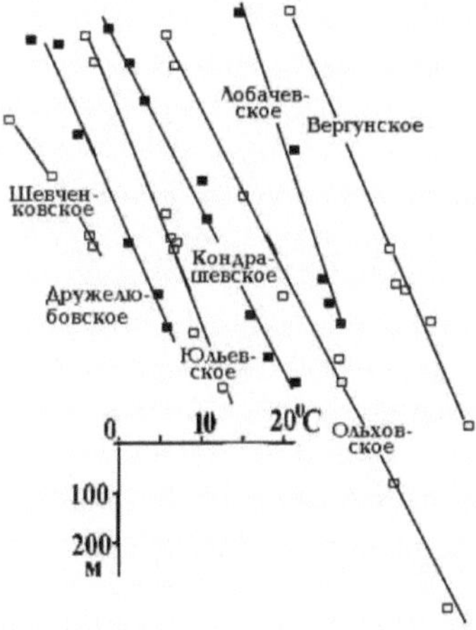

Fig. 1.2. Results of T determinations in wells of several fields of the northern

edge of Donbass and the adjacent part of the Voronezh massif slope.

Obviously, the distribution is quite symmetrical, and typical discrepancies are only 2.5^0 C, which is noticeably smaller than their values due to the usual errors of downhole temperatures used earlier for TP calculation [16]. Thus, the application of the new version of the methodology promises not only the absence of additional distortions of natural T, but also their more accurate determination.

Replacement of drilling fluid in the well during formation water tests can contribute to the determination of T, which is not distorted by the drilling process. This assumption has been tested on many examples that confirmed its validity.

Some data of this kind can also be given for Donbas (Fig. 1.2). Obviously, the deviations from the averaging straight lines are on average about 1^0 C. It is essentially less than at similar constructions with the use of bottomhole temperatures (3^0 C), and it is necessary to take into account the fact of measurements in different wells in each field, while bottomhole T was taken in one well. The linear variation of T with depth indicates that the thermal conductivity of rocks is stable in large depth intervals. Comparison of the results of TP calculation in the repeatedly studied wells by the considered temperatures leads to the conclusion that they agree with those calculated by downhole T. The insignificant difference (on average - 3 mW/m excess of the new values over the old ones[2]) may indicate the heating of productive horizons by relatively recently introduced fluids.

1.2. Thermal conductivity of rocks in Ukraine

The value of thermal conductivity of rocks from different regions of Ukraine has been studied (including - with the participation of the authors) in different years by different methods by different specialists [17, 20, 29, etc.]. We will not dwell on the technique and methodology of these studies, which are described in detail in these works. We only note that for rocks with large porosity (mainly sedimentary rocks), λ values with natural moisture content were used to calculate TP.

Ukrainian Shield (USh). About 1200-1300 samples have been studied on its territory (Table 1.1).

Table 1.1. Thermal conductivity of crystalline rocks of the Ukrainian Shield (in W/m·^{0} C)

Breed	Control room units					
	А	Б	В	Г	Д	Е
Granite	2,7	2,9	3	2,5	3	
Gneiss and crystalline shale	2,6	2,5	2,5	2,9	2,9	2,3
Migmatitis and rapakiwi	2,2					
Charnockite	2,1	2,6				
Gabbro, anorthosite	1,7	2,1			2,5	
Pyroxenite	1,8	2,5				
Albitis	2,6					
Enderbit		2,4				
Amphibolite		2,4		2,2		
Quartzite, hornblende, dolomite, marble		4,2	3,7	3,7	3	
Basalt			2,5			
syenite	2					
Serpentinite		2,1				
Diorite						1,7

Calcifer		2,3				
Orthoclasite			2,8			

The shallow sedimentary cover of the shield and its slopes has been studied in relative detail only in some areas. In total, about 200 samples have been studied [18-20, 29, etc.].

In the central part of the shield (Kirovograd block) the rocks of the cover and weathering crust with total thickness from 30 to 130 m have been studied. Excluding the lower depth interval, where weathering crust and fragments of crystalline basement formations play a significant role, the average value of thermal conductivity is about 1.7 W/m·^{0} C. In the weathering crust, thermal conductivity increases up to 1.9-2 W/m·^{0} C, and as it becomes enriched with debris, it increases even more.

In the Boltysh depression located to the north, the thickness of the cover is measured in hundreds of meters. Here, the thermal conductivity of Meso-Cenozoic rocks has been studied, which does not differ significantly from that established for similar formations of the Dnieper-Donets depression -1.6 W/m·^{0} C. The thermal conductivity of combustible shales and Cenozoic basalts encountered in the depression is lower - 1.2-1.5 W/m·^{0} C.

The thermal conductivity of Neogene-Quaternary and Paleogene rocks of the upper part of the section was studied on the northwestern, northeastern, and southern slopes of the shield. It ranges from 1.2 to 1.7 W/m·^{0} C and mainly depends on their porosity and natural humidity. On the western slope, a layer of chalk (Cretaceous age) with high (about 2.1 W/m·^{0} C) thermal conductivity appears in the section.

Dnieper-Donets depression. In total, about 1300 samples were studied [2, 29, etc.]. Hereinafter, the numbers in parentheses are the number of samples in rock groups.

Cenozoic-Cretaceous sands and sandstones (12) have an average λ of 1.65

W/m$^{.0}$ C. According to [29] Mesozoic sandstones (10) - 2.1 W/m^0 C. Mesozoic clays (23) are 0.87 without hydration and 1.65 W/m$^{.0}$ C with hydration. Another series of determinations (30) - average λ about 1.4 W/m$^{.0}$ C. According to [29] (35) - about 1.35 W/m$^{.0}$ C, with humidification clearly insufficient. Cenozoic terrigenous rocks in general (clays, loams, loams, sand moistened) (40) - 1.65. W/m$^{.0}$ C.

Average values for stratigraphic differences are: Cenozoic - 1.65, Cretaceous - 1.85, Jurassic - 1.5, Triassic - 1.65 W/m$^{.0}$ C.

The λ of Carboniferous rocks have been studied in particular detail (about 1000 samples) [2, 29, etc.]. Histograms were constructed for mudstones, siltstones, and sandstones, with several hundred samples for each rock type. As a result, the average effective thermal conductivity of Carboniferous rocks (a mixture of 80% mudstones and siltstones, 20% sandstones, and an insignificant amount of limestone) was found to be 1.9 W/m$^{.0}$ C. The value of thermal conductivity of salt (the salt itself has λ about 5-6 W/m$^{.0}$ C) taking into account its "contamination" with terrigenous rocks is about 4.1 W/m$^{.0}$ C.

Less studied rocks of Permian (data on numerous small salt layers in the rocks are included) and Terrigenous Devonian - 2.25 and 2.5 W/m$^{.0}$ C, respectively.

Donbass. In the region, the maximum number of samples for Ukraine was studied (mainly by production organizations of the Ministry of Geology of Ukraine) - about 5000 [2, 29, etc.]. However, some of them were studied using substandard methods and equipment (as a result, values sharply differing from the rest and from each other were obtained), so about 4000 determinations were involved in the consideration. Naturally, the maximum amount of data characterizes the Carboniferous rocks. Histograms of λ distribution of the main varieties of mudstones, siltstones, sandstones, and limestones from different parts of the region were constructed for them. The effective average λ is about 2 W/m$^{.0}$ C. However, values calculated for localized areas of Donbass were used for TP calculations, in determining which the relative amount of sandstones within the section penetrated by the boreholes was taken into account. The anisotropy λ of Carboniferous rocks within the Main anticline with steep bed dips was also taken into account.

The thermal conductivity of rocks of the **Meso-Cenozoic** cover and Devonian in Donbas does not noticeably differ from that established in the DDV. The value of λ of Permian rocks is slightly higher - 2.4-2.5 W/m·^{0} C

Crimea and the northern shelf of the Black Sea [2, 29, etc.]. In total, about 600 samples were studied. Tertiary age terrigenous rocks were studied in comparative detail (100), and a histogram of λ distribution was constructed. A modal value of 1.5 W/m·^{0} C was obtained.

Cretaceous limestones were studied on 200 samples, a histogram was constructed, the modal value of thermal conductivity is 2.3 W/m·^{0} C. Tertiary limestones (taking into account porosity and moisture content) (50) - 2.3 W/m·^{0} C practically do not differ from them. Limestones of the Paleozoic basement are poorly studied (10) - 2.8 W/m·^{0} C. Cretaceous and Tertiary marls (60) are characterized by an average λ value of 1.9 W/m·^{0} C.

Argillites and siltstones of the Cretaceous (60) - average value of about 2 W/m·^{0} C. Argillites, siltstones and sandstones of the Jurassic, Triassic and Paleozoic (60) - 2.5 W/m·^{0} C.

Tuffs of Cretaceous age were investigated on a small number of samples (10) - average λ=2.1 W/m·^{0} C.

The upper part of the section on the shelf is represented by unconsolidated silts with very low thermal conductivity - about 0.8-1.2 W/m·^{0} C. They are studied in the form of samples taken from soil-bearing tubes with preservation of natural moisture. In total, about 50 samples were studied.

South Ukrainian monocline. In this region, the thermal conductivity of rocks is poorly studied (less than 100 samples). In the Cenozoic and Mesozoic (Cretaceous) parts of the section, the λ values do not differ from those obtained in the Crimea and on the southern slope of the Ukrainian Shield (see above). The Paleozoic rocks (reached by drilling in the Prydobrudzha Trough) have significantly higher thermal conductivity - 2.2-2.6 W/m$^{.0}$ C.

Volyno-Podolsk plate and the Lviv Paleozoic trough. Here, samples of the Meso-Cenozoic cover (100), whose thermal conductivity does not differ from that established on the western slope of the Ukrainian Shield (see above), have been studied in relative detail. The thermal conductivity of Paleozoic terrigenous rocks was studied on about 30 samples. It averages 2.35 W/m$^{.0}$ C [29 et al.].

The thermal conductivity of Paleozoic limestones and dolomites (30) is somewhat higher, averaging 2.75 W/m$^{.0}$ C.

Carpathian region. Its study of thermal conductivity is extremely uneven [2, 29, etc.]: the rocks of the Precarpathian and Transcarpathian troughs are mainly studied, while the folded Carpathians are represented by a limited number of samples from several boreholes. In total, about 250 λ determinations were made.

The average thermal conductivity of poorly compacted clays, mudstones, siltstones and (in small quantities) limestones of the Predkarpattya and Zakarpattya troughs (60) is 1.8 W/m$^{.0}$ C. Compacted limestones, mudstones, shales, and sandstones of the Meso-Cenozoic of the troughs have an average λ = 2.4 W/m$^{.0}$ C (30). Dense sandstones of the Cretaceous-Paleogene flysch in the area of the Folded Carpathians thrust on the Precarpathian Trough: λ = 2.6 W/m$^{.0}$ C (50). Volcanogenic rocks of the Neogene of Transcarpathia (60) - 1.9 W/m$^{.0}$ C.

In the Folded Carpathians, clayey varieties of flysch rocks are characterized by an average thermal conductivity value of about 2.3, sandstones - up to 3.4 W/m$^{.0}$ C (60).

1.3. Change in thermal conductivity of rocks under the influence of temperature and pressure

The necessity of knowledge of thermal conductivity of deep crustal and upper mantle rocks for TP and T calculations is obvious. Such information is mainly needed for the construction of deep thermal models. Therefore, the consideration of this issue in this chapter is somewhat artificial, since the relationships revealed were not much used to estimate the thermal conductivities used in borehole TP calculations. An exception in principle could be the calculations in some areas of the Folded Carpathians and the USH.

At present, some material has already been accumulated on changes in λ of rocks as a function of composition, pressure (p), and temperature. Below we attempt to summarize these data for crustal layers and mantle tops for different PT conditions. Literature information [29-32, 35, 37, 39, 41, 45-47, 49, 50, 52, etc.] and some of the authors' own results are used.

The data on the following layers were analyzed: sedimentary (λ_1), volcanogenic-sedimentary (λ_2), granitic (λ_3), transitional (λ_4), basaltic (λ_5), and upper mantle rocks (λ_6). The ideas about the composition of these horizons are based on the works [3, 4, 8-10, etc.].

The material collected for each of the layers is quite different in terms of completeness of collections, temperature and baric ranges of studies. Single experiments, which make it possible to consider independently the effects of T and P, show that the summation of the effects of T at normal P and the effects of P at normal T does not fully correspond to the results obtained with a coordinated gradual change of T and P, i.e., the effects are not additive. However, at present it is possible to take into account both factors only using data from separate experiments.

The relationship between λ and P even at normal temperature is still poorly understood, so the conclusions below may be revised in the future.

The well-known significant influence of temperature on thermal conductivity requires preliminary determination of the thermal regime of the subsurface, for which the variation of λ with depth is calculated. It is desirable to obtain T distribution without geothermal calculations, which inevitably include thermal properties of the medium. Such an opportunity is provided by geothermometer data [6, etc.]. At present, they are obtained on all continents except Antarctica and are sufficiently representative. Among the identified varieties of thermal regimes there are two polar ones: cold platform and hot geosynclinal (protogeosynclinal in the Precambrian). The high-temperature variant is probably predetermined: it achieves partial melting temperatures in the middle part of the crust for rocks of amphibolite facies of metamorphism, in the lower part for basic granulites, and in the upper mantle for mantle pyrolite (Fig. 1.3). Below we will consider changes in thermal conductivity with depth for both variants of T distribution. The pressure in all cases was considered to be lithostatic, and in the crust it increased with depth in accordance with its average density - 2.8 g/cm^3 , in the mantle - 3.33 g/cm^3 (the crust thickness was assumed to be average according to GSZ data for many regions - 42 km).

Influence of pressure. The authors know the results of only about 20 experiments in which the thermal conductivity was determined in the pressure range up to 1 GPa and more. The data on about 50 experiments for relatively low pressures - up to 0.30.6 GPa - are also known. Samples of granites, charnokites, gneisses, pyroxenites, and forsterite have been studied. The changes in λ are quite different, and it is difficult to determine whether they are affected by individual features of the studied formations or reveal characteristic features of large classes of rocks. Therefore, it was decided to limit ourselves at the achieved level of research to the construction of a general dependence of the relative change of λ on pressure. It averages all known data (except for information on the monomineral forsterite aggregate, for which $\lambda r/\lambda_0$ increases slightly more - up to 1.33 λ_0 per 1 GPa) and was used to account for pressure in all layers.

The obtained dependence can be represented as a direct relation $\lambda r/\lambda_0$ with depth (H - in 10 km):

$\lambda_p = \lambda_0 (1.04 + 0.06N - 0.005N)^2$

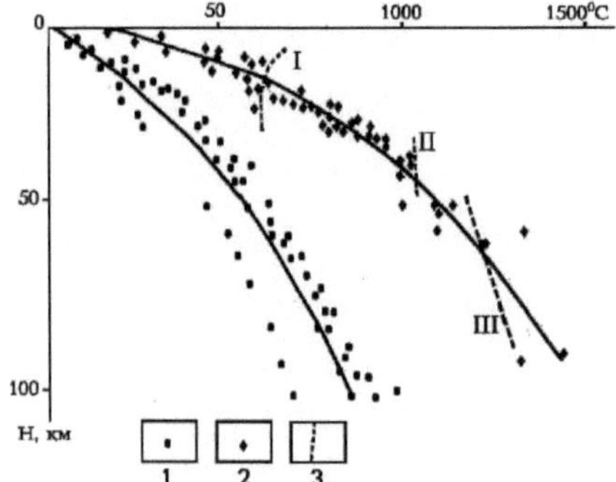

Fig. 1.3. Temperature distributions from geothermometer data under the platform (1) and alpine geosyncline (2), compared with solidus temperatures of crustal (I and II) and mantle (III) rocks.

It should be noted that the variation of $\lambda p/\lambda_0$ in the depth interval 0-5 km does not obey the above formula.

It is significantly more intense. However, in most real situations, it can be assumed that for such shallow depths λ differs little from that obtained under normal conditions. If necessary (arising in the case of anomalously high geothermal gradient), the pressure effect at depths of 0-5 km can be taken into account in the form $\lambda_p = \lambda_0 (1 + 0.14H)$

Influence of temperature. Thermal conductivity at different T and atmospheric pressure was determined for about 250 samples of rocks of different composition, taken in different regions of the CIS and other countries, in the temperature ranges from 20-200 to 20-1200^0 C. For some of the studied rocks solidus temperatures were reached.

Table 1.2. Accepted PT-conditions of the lithosphere and the relation of λ to P

Depth, km	Pressure, GPa	$\lambda p/\lambda 0$	Temperature, °C	
			Minimum	Maximum
5	0,14	1,07	80	260
10	0,28	1,10	160	480
15	0,42	1,12	240	640
20	0,56	1,14	300	740
30	0,84	1,17	420	910
40	1,12	1,20	520	1040
50	1,43	1,21	600	1150
60	1,76	1,22	670	1240
70	2,09	1,23	720	1320

The relationship between thermal conductivity and temperature is complex; when a certain degree of heating is reached in most rocks, thermal conductivity decreases and then increases. However, the PT-conditions under both variants of depth temperature distribution for rocks of most layers are such that the effect of λ growth with increasing T is not achieved. Therefore, such an important feature of thermal conductivity behavior is not considered below.

The processing of experimental results to obtain regional and generalized dependences of λ on T for crustal layers involves a number of difficulties caused by incomplete information. Not all its parts (concerning high and low thermal conductivities in the sample for a layer, high and low temperatures) are equally represented. As a result, the processing assuming equal weight of all points in the coordinate system λ and T can lead to distortion of real dependences. Practically, this processing difficulty was expressed as follows. The lowest λ was obtained for samples from Siberia. For the same data, the maximum temperature range (more than 600-800°C) was investigated. In the region of high T, where data for other regions are not available, the information on Siberia becomes decisive; the generalization curve for the layer approaches the regional one for Siberia. It turns out that the shape of the generalization curve differs from the shape of all regional curves; it demonstrates a sharper change in λ with T. This effect is especially pronounced in the construction of generalization curves for the granite and transition layers. Here, a correction that takes into account the shape of the regional curves seems logical.

Less intensive distortions of generalization curves for other layers during the transition from low T to high T are also probable. Therefore, it should be considered that the errors in the construction of generalization curves reach tenths of W/m$\cdot^0$ C and the details of their shape cannot be studied from the available material. Thus, it is possible to use simplified types of the relationship between λ and T.

Regional curves are more reliable in this sense, they should be approximated by functions that describe all identified details of the curve shape.

Model of thermal conductivity of the crust and upper mantle. The results of λ calculation, taking into account the influence of T and p, are presented in Fig. 1.4 for both variants of temperature distribution with depth. The following distribution of crustal layers is conventionally accepted: 0-5 km - sedimentary, 5-10 km - volcanogenic-sedimentary, 10-20 km - granitic, 20-30 km - transitional, 30-42 km - basaltic. We can see a sharp difference in the behavior of λ with depth for temperature models, as well as a stepwise increase of λ with depth in the consolidated crust. Probably, in natural conditions, the change of composition does not occur so sharply and deep changes in thermal conductivity are smoother.

It is useful to determine the average characteristics of the crust and mantle tops. They were calculated as follows: $\lambda_\Sigma = (\Sigma\ (N_i/\lambda_i\ ))^{-1}$, where Ni *is the* relative thickness of the layer in the pack. We obtain $\lambda_{1\text{-}5}$ =2.3 and 1.95 W/m .0 C for the two thermal models, λ_{1-6} =2.5 and 2.2, respectively, when mantle rocks up to 70 km are included.

It should be noted that in the case of the high-temperature variant of the section, the convective heat conduction (which has only been considered so far) is supplemented by convective heat conduction in the layers of partial melting. Their location in the section and their power are clear from Fig. 1.3. The evaluation of the Nusselt criterion for probable convection conditions (we emphasize that we are talking about the evaluation, for a more reasonable calculation it is necessary to consider the convective process in detail) allows us to determine the growth of the effective thermal conductivity by a factor of 1.7 compared to the conductive one.

In the melted parts of the granite and transition layers, λ reaches 2.6 and 3.5 W/m$\cdot^0$ C, respectively, and in the partially melted mantle layer (up to 70 km) 4.5 W/m^0 C. The influence of the low-power layer of partial melting in the "basalt" can probably be disregarded in the estimative nature of the calculations. With the convective component, the effective thermal conductivity of the crust in the hot variant will be the same 2.3 W/m$\cdot^0$ C as in the cold variant. The thermal conductivity of the upper mantle horizons is 3.3 W/m$\cdot^0$ C, and of the whole depth interval under consideration (0-70 km) - 2.6 W/m$\cdot^0$ C.

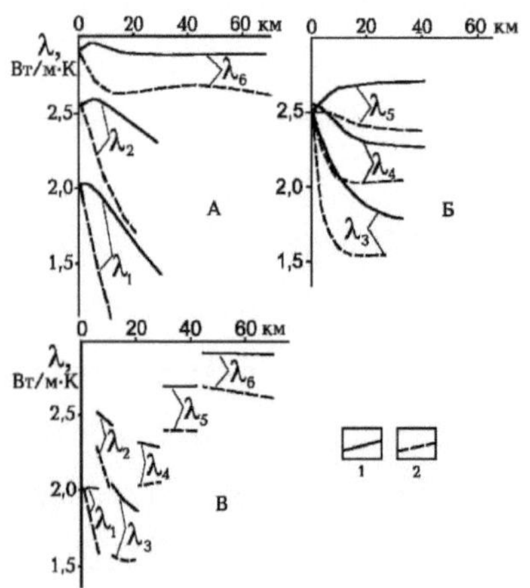

Fig. 1.4. Distribution of thermal conductivity in the tectonosphere layers without taking into account (A, B) and with taking into account (C) the power of layers.

1 - cold variant, 2 - hot variant.

It does not follow from the obtained coincidence of the mean λ. for the hot and cold variants that all intermediate distributions of T will lead to the same results.

Some temperature decrease compared to the adopted hot variant will lead to disappearance of partial melting zones, i.e. convective component.

There will be no corresponding increase in the conductive component, as T will remain still much higher than in the cold variant. Therefore, a noticeable decrease in the average thermal conductivity of rocks at depths of 0-70 km is likely. Naturally, the values of λ. will still be higher than for the hot variant without taking into account the convective component. Therefore, we can expect λ of the crust to decrease to 2.0-2.1 W/m^0 C, of the mantle - to 2.7-2.8 W/m$^{\cdot 0}$ C, and of the average value in the 0-70 km interval - to 2.3 W/m$^{\cdot 0}$ C. Probably, these differences from the above estimates do not significantly exceed the accuracy of calculations.

The calculated values of λ can be subjected to control by independent data on the heat flux and crustal heat generation. Such a check seems reliable only for the crust, where the nonstationary component in both variants can be considered not very significant (in the geosyncline, its value changes markedly during 20-30 Ma near the maximum of the heat flux; it was averaged because the geothermometer data averaged the temperatures of this period).

The average thermal conductivity of the crust is calculated by dividing the average heat flux by the average geothermal gradient (determined by the T difference between the roof and the basement). The latter is 1.275 and 2.575^0 C/km for cold and hot variants. The average heat flux on the Precambrian platform is 45 mW/m^2 , in the Alpine geosyncline - 75 mW/m^2 . The part of it produced by radioactive decay in the crust of our adopted structure is about 30 mW/m^2 . Thus, the average flux in the crust is 30 and 60 mW/m^2 , which corresponds to an average thermal conductivity of 2.35 W/m$\cdot^0$ C. Obviously, the test was successful.

In conclusion, it should be noted that the task of determining λ of crustal and upper mantle rocks under RT conditions in the subsurface cannot be considered 26

finally solved. The single experiments on the coordinated effect on the thermal conductivity of p and T mentioned at the beginning (the geothermal gradient adopted in the experiments approximately corresponded to the cold scenario considered by us) demonstrate significantly larger λ for rocks of the granite layer than those obtained by us [30, 31]. At close values of λ_0 depth values are 1.5 times higher, there is a *30%* increase in thermal conductivity, and according to our calculations, a 20% decrease. On the other hand, data on forsterite for the same variant of the temperature section reveal a practical invariance of λ at depths of about 36-60 km compared to λ_0 , which agrees well with our data for mantle rocks. For forsterite, a hot variant is also possible, but only under crustal RT conditions. The observed reduction in λ_0 of the order of 10-15% agrees well with our calculations. In general, the work done has allowed us to outline the main regularities of changes in the thermal conductivity of crustal and upper mantle rocks at the achieved level of their study. The refinement will be associated primarily with the development of experiments in which the samples are subjected to a coordinated effect of pressure and temperature.

Note that the volumetric heat capacity of lithospheric rocks, according to the few data available, increases from the upper part of the crust (consolidated sediments) to the middle part from 2.5 to $4.2 \cdot 10^6$ $J/^0 C \cdot m^3$ and remains at this level in the upper mantle horizons. Accordingly, the average thermal diffusivity can be estimated at about $7 \cdot 10^{-7}$ m /s.^{2}

For the rocks of the consolidated part of the crust of Ukraine, individual estimates of thermal conductivity depending on depth [30-32 and others] can be more or less reasonably made only for the granite layer. Probably, for its entire thickness (0-15 km) we should assume λ = 2.65 $W/m \cdot ^0 C$ for the cold variant of the section. For the hot variant, λ will decrease by about 0.4 $W/m \cdot ^0 C$ to the bottom of the layer. The rocks of other layers are represented by a small amount of data.

1.4. Calculation of heat flow

Numerous works, including those of the authors [17, 19, 29, etc.], have been devoted to the calculation of TP based on the results of determining the geothermal gradient and thermal conductivity of rocks in individual depth intervals traversed by wells. The technique of these calculations in case of its use on the territory of Ukraine did not differ from the generally accepted one. When determining the TP using the average gradient between the bottom of the well and the surface, the accuracy of knowledge of the surface (mean annual) temperature plays a significant role.

The latter was established based on data from a network of meteorological stations, which made it possible to determine T0 over the entire territory, but only for one type of surface cover and with a small accuracy - about $0.51.0^0$ C. When working with downhole T data in deep wells (1000 m and more) such an error is acceptable (in many cases it turns out to be less than the downhole temperature measurement error - see above). The real error in determining the geothermal gradient in deep wells does not exceed a few percent on average.

For shallow wells (several tens of meters deep) more precise information is needed. It is obtained from the results of special studies in shallow wells of the corresponding regions of Ukraine. In them, measurements were made at depths below the layer of seasonal T fluctuations and the obtained depth distribution was extrapolated to the surface, if necessary, taking into account changes in thermal conductivity.

The most complete example of using this approach is the study of TP in the central part of the Ukrainian Shield, where bottomhole temperatures were set in 1500 wells 30-130 m deep. Based on meteorological data within the region bounded by coordinates 49°00-47°40' N and 32°00 -33°20' E, a rather simple type of "global" trend of T0 was established (Fig. 1.5): To = 9 - k ((f - 49) - 0.25 (d - 32)), where f and d are latitude and

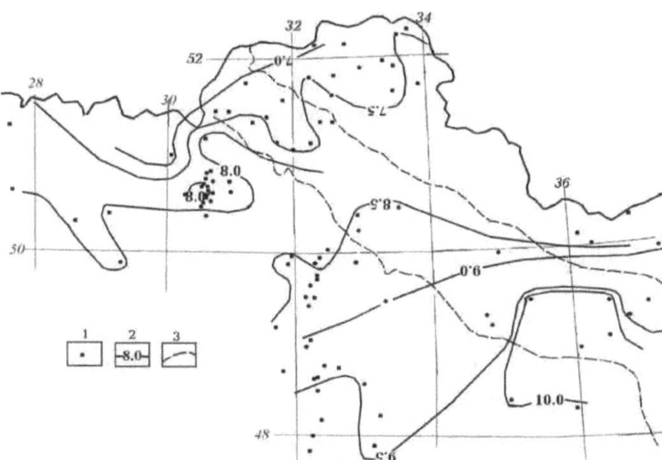

Figure 1.5: Surface temperature distribution in northwestern Ukraine.

1 - definition points, 2 - isolines T_0 , 3 - DDS contour.

longitude of the T_0 point, k - 1°C/°coord . The average deviation of T0 values at the meteorological stations (available only in the northern part of the area shown in the figure) from the global trend is about 0.05°C.

The obtained distribution was controlled and supplemented with well data. Dependences of bottomhole T on the measurement depth were plotted for 30 sites. In order to obtain each of them, measurements at many dozens of points were used. The T0 determined by extrapolation to the surface (using the least squares method) confirm the global trend in the northern part of the region and do not correspond to it in the southern part. Probably, in the latter case we are dealing with the regional T0 field; it should be constructed and used separately. The average values of T0 were taken at the border of the two zones (see Fig. 1.5).

When extrapolating the distributions of downhole T at individual sites to the surface, local TQ anomalies were also detected in the northern part of the region (see Fig. 1.5). The latter are related to the long-established fact of T0 lowering in the forest. These perturbations were taken into account in further calculations. It is possible that similar anomalies are also present in the areas where it was not possible to construct a qualitative distribution of downhole temperatures by depth. Then T0 will be determined with a large error, but such a danger exists only in a limited number of points and is partially eliminated in subsequent calculations.

The effective average thermal conductivity of rocks was used to calculate the average gradient in the depth interval from bottomhole to surface. It was calculated from the description of the borehole section, and the average λ values established for these rocks in the region were assigned to lithologic and stratigraphic differences. The effectiveness of this approach was substantiated by comparing the results for one of the most studied regions of Ukraine in terms of thermal conductivity - the Dnieper-Donets depression [19]. In several dozens of wells, where the temperature distribution along the borehole was determined after sufficient settling and the thermal conductivity was determined by the maximum number of samples, the TP was calculated using the "local" λ and the regional average. In the latter case, the convergence of interval results doubled, and the average values practically did not differ. According to these data, we can state that the error in the value of TP, introduced by inaccurate determination of thermal conductivity, does not exceed a few percent in real experimental conditions in sedimentary basins of Ukraine. For studies in boreholes that have opened crystalline rocks, the error is clearly higher and it is difficult to estimate it. This remark does not apply to the overwhelming majority of TP determinations on the Ukrainian Shield [15]: here, the rocks of low thickness sedimentary layer played the main role in calculating the average effective λ.

There is no doubt that in principle there is a problem of taking into account the influence of natural conditions (temperature and pressure) on the value of λ at real depths of geothermal studies in boreholes. It concerns rocks of sedimentary, sedimentary-volcanogenic and granitic layers. Noticeable effects (change by 10% or more) are possible for the hot variant of the section at depths greater than 3 km (Fig. 1.4). In the first two layers from above, the effect under consideration is superimposed on the intensive growth of rock thermal conductivity with depth due to an increase in the degree of catagenesis. The visibly compacted and mineralized sediments corresponding to the depths of 3-6 km (effusives do not play a significant role in the composition of the volcanogenic-sedimentary layer at the considered depths in the conditions of Ukraine) are less affected by PT conditions than loose ones. In any case, the corresponding effect was not recorded in the form of interval changes in TP in the only sedimentary basin with a hot section and sufficient depth of the studied wells in the study area - in the Folded Carpathians.

Observations in the granite layer at the specified depths were made in only one well in Ukraine - Krivoy Rog superdeep well - obviously in the conditions of cold section.

The data presented in the chapter allow us to assert that the techniques used and the experimental material involved make it possible to achieve the usual in the world geothermia error in determining the observed value of TP - about 5-10%. This conclusion is confirmed by comparing the results obtained in several dozens of wells in all major regions of Ukraine, where the heat flux was determined using different techniques.

1.5. Introduction of heat flux corrections

The calculated value of TP is subject to numerous distortions under the influence of processes in the near-surface zone. However, for building thermal models of the subsurface, a corrected flux, which we will call depth flux - GTP (hereinafter we will refer to this parameter), is necessary. Its determination requires the introduction of numerous corrections, some of which are briefly discussed below. More complete information is given in [16 and others].

1. First of all, it is a correction for paleoclimate, i.e. for the difference between the past surface temperature and the present one. Naturally, its value changes with the depth and length of the period of accounting for the climate of past epochs. Practically, for the deepest boreholes (up to 6 km) it is enough to take into account the variations in surface temperature over the last 1.2 million years. The correctness of the correction was checked on the shield, where the influence of groundwater overflows is insignificant. With correct correction, the value of TP at all depths should be constant within the error. This result has been achieved. It is shown that corrections in shallow wells can change the TP by tens of percent.

2. Hydrogeological effects are represented by a whole family of corrections that take into account heat transfer in the near-surface layer of the Earth's crust by water movement, including a vertical component. These include surface temperature anomalies (in particular, due to increased evaporation from shallow groundwater mirror), which shift the entire thermogram during long-term existence. Significant distortions can be associated with meteoric water

infiltration. In shallow wells they reach 20-70% of TP.

The gradient and water transfers between low-pressure aquifers, which, as a rule, have seasonal character, influence the gradient. In areas of intensive exploitation of aquifers, the influence of pumping (formation of depressions) can be detected. In mountainous areas intensive temperature decrease (up to the formation of a depth interval with a negative gradient) in the beds of underground rivers is very clearly manifested. Less intense, but still sometimes quite noticeable effect of groundwater movement through the aquifers under the influence of anomalously high reservoir pressures. Very strong local anomalies occur when water flows through permeable fault zones.

All of these influences can be eliminated (if any of them are not attributed to the studied TP anomalies) only by having information about the corresponding processes and developed calculation methods. The latter are summarized in the works of the authors [16, etc.].

3. The structural effect, which is a concentration of TP in blocks of rocks of increased thermal conductivity, is quite widespread in areas with non-zero-rizonal position of crustal layers. It reaches significant values (first tens of percent) only in pronounced folds (Main anticline of Donbass, etc.). More often the value of the calculated corrections does not exceed the usual error of heat flux determination.

4. In the Folded Carpathians and on their border with the Pre-Carpathian Trough, the effect of young thrusts is noticeable. It is created by a rock massif with a vertical temperature gradient common for the period preceding the thrust, placed on top of an autochthonous massif, which at the first moment preserved the same T distribution. In other regions such young transformations are absent, and no corrections are introduced.

5. Quite significant values can be reached by the influence of young rapid sedimentation, which makes sense to consider when measuring TP through the seafloor with a buried probe (i.e., when determining the geothermal

gradient on the basis of the first meters). Practically, we are talking about the Black Sea, where to this distortion may be added the influence of degradation of the gas hydrate layer at a shallow depth below the bottom surface [5].

Chapter 2: Depth heat flux distribution

2.1. General characterization of the study

The distribution of TPs within the territory of Ukraine is considered in accordance with the scheme of tectonic

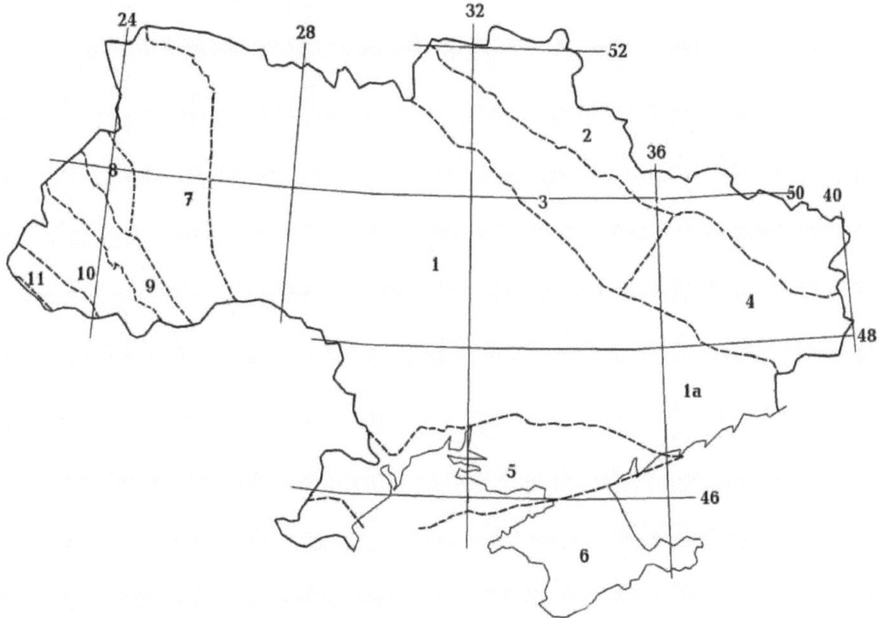

Fig. 2.1. Main tectonic units of the territory of Ukraine.

zoning shown in Fig. 2.1. It shows the massifs within which geosynclinal activation has not occurred since the Precambrian: Ukrainian Shield and its slopes (1), including the Azov Massif (1a), Voronezh Massif and its slope (2), Volyno-Podolsk Plate (7) and South Ukrainian Monocline (5), Hercynian rift of the Dnieper-Donets Basin (3) and Hercynian parageosyncline of Donbass (4), a small fragment of the Hercynian-Caledonian West European Plate (8), the Epikimmerian Scythian Plate (6), and the Alpine geosyncline of the Eastern Carpathians, including the Precarpathian Trough (9), the Folded Carpathians (10), and the Transcarpathian Trough (11).

The density of TP determinations on the territory of Ukraine is extremely uneven. A significant part of the UCH and its slopes, as well as the slopes of the Voronezh massif are poorly studied. It is difficult to increase the network density in these regions, because thermometry is not included in the GIS complex, and there are few wells suitable for measurements. But it is necessary to do it, because there is no reason to consider the deep heat flow of the shield, massif and their slopes as low and sustained (as it was generally recognized recently). In the detailed investigated areas of the Shield, Kirovograd and Dnieper anomalies of complex shape were found, within which TP reaches 70-75 mW/m^2 with a background value of 45 mW/m^2 . Thus, the priority direction of future research is obvious.

Virtually all of the TP values below have been published.

Ukraine (area - 600 thousand km^2) surpasses all countries of comparable or larger size in terms of average density of heat flux determination network. Here it is installed in more than 13000 wells (about 20 determinations per 1000 km^2). For comparison: in the rest of Europe (on the area of about 10 million km^2) - about 4000 determinations.

Single determinations are grouped in points differing in coordinates by 1' or more. Each of them contains from 1 to 30 single determinations. The total number of determination points on the territory of Ukraine and the Black Sea-Azov shelf is 5700. In trapezoids of 20' latitude x 30' longitude (37x37 km) - there are only 500 of them (since the map covers the shelf and small fragments of the territories of neighboring countries adjacent to Ukraine) - from 0 to 258 points. There are no definitions in 33% of trapezoids. This is mainly the territory of the Ukrainian Shield, its slopes, slopes of the Voronezh Massif, the southern part of the Folded Carpathians, and the shelf water area. In the rest of the territory the study is extremely uneven. About half of the determinations were made in Donbas.

The very procedure of heat flow mapping seems to be non-trivial at present.

The fact is that due to the usual poor study of the TP, this issue in the geothermal literature has simply not been considered (except for some works of authors). The available "maps" of heat flow of different regions, in fact, do not correspond to this name. When working with relatively dense observation networks on the territory of Ukraine, it was necessary to formulate at least elementary requirements for the construction of maps and only then proceed to their actual production and field description.

In geologic mapping, which inevitably uses an irregular network of observations (outcrops, boreholes, etc.), the minimum requirement for its density is the presence of 1 observation per 1 cm^2 map. Attempts to justify the network density for regional TP maps lead to approximately the same values. For the territory of Europe it turns out to be possible to construct only a map with a scale of 1 : 5 000 000 (except for some regions). For Ukraine the average scale is defined as 1 : 700 000. Taking into account that half of the points are concentrated in Donbas, which occupies less than 10% of the territory of Ukraine, for this region it is possible to allow mapping at a scale of 1 : 200 000 (for its separate regions - 1 : 50 000 - 1 : 25 000), for the rest of the territory of Ukraine - 1 : 1 000 000. Quite a wide distribution of areas where the network density outside Donbas is much lower than average leads to the necessity to reduce the map scale to 1 : 2,500,000, while some of the heat field features in the most studied areas cannot be shown.

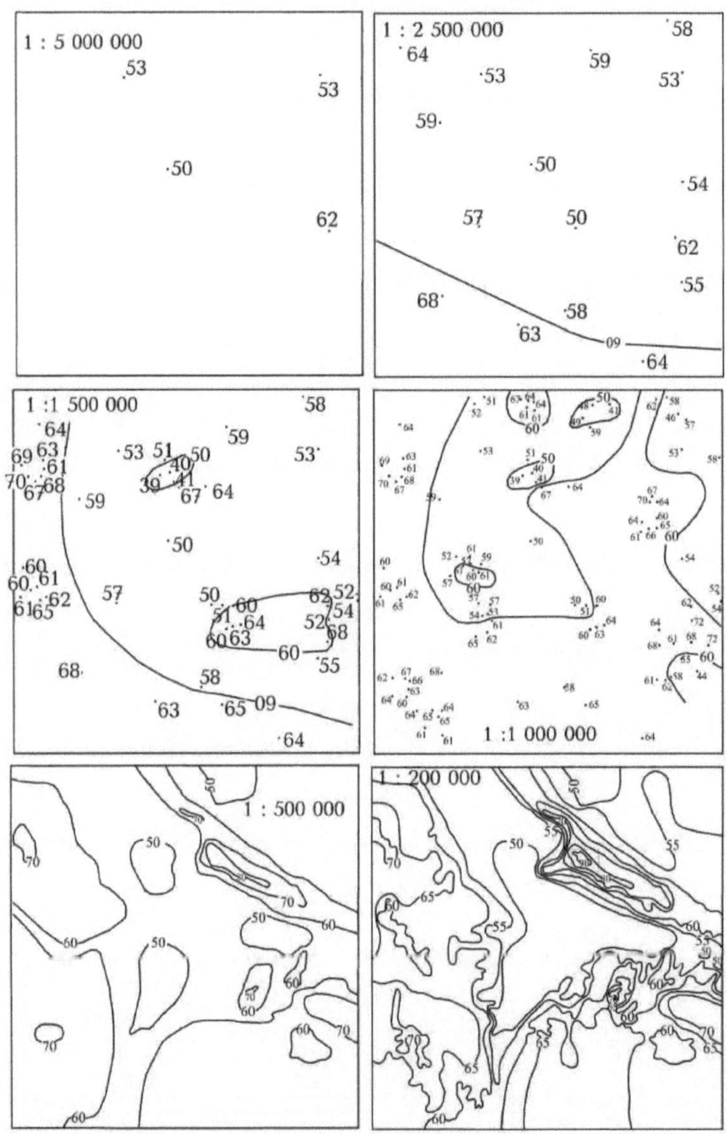

Fig. 2.2. Multiscale maps of deep TP in the central part of Donbass on the territory of 100x100 km.

Therefore, local maps of different scales for separate regions or their parts are given below. On all maps points are points of single or group TP determinations, isolines are in mW/m .^{2}

The study of the structure of the TP field in Ukraine allows in most regions to statistically reliably distinguish the background and anomalous data sets and to estimate the standard deviations from the modal values. They appear at the level of several mW/m^{2} , i.e., quite comparable with the error of TP determination. Therefore, the isolines of the depth heat flux, conducted with a step greater than twice the error, in most regions should differ by 10 mW/m^{2} . In this case, they will reliably reflect the TP anomalies. In some regions with low TP or particularly dense network and low relative error, it is possible to draw auxiliary isolines at 5 mW/m^{2} . In the Transcarpathian sag with the highest TP values, the step of isolines was doubled (isolines of 80, 100 and 120 mW/m were drawn).^{2}

The map should reflect the regional features of the thermal field at the achieved level of study. There are significant areas within Ukraine where the average network density meets only the scale of 1 : 5 000 000. On the rest of the territory more detailed maps can be constructed. In this regard, let us consider the issue of the scale (network density), which is sufficient to identify regional features of TP distribution. It is convenient to do it on the example of a part of Donbass, where the current network density is obviously sufficient for solving the task at hand.

Fig. 2.2 shows 6 maps reflecting the real stages of study of a part of the region with the size of 100x100 km during 40 years. At a scale of 1 : 5 000 000 (4 TP definitions) it is impossible to speak about any structure of the polygon field, only an estimation of the average TP value is possible, but its reliability is unclear. At a scale of 1 : 2 500 000 000 (16 determinations) the first sign of TP difference in parts of the polygon appears: the heat flux growth in its south-western fragment. At a scale of 1 : 1,500,000 (44 determinations), additional elements of regional TP change are detected, in addition to an increase in relative decrease. At a scale of 1 : 1 000 000 (100 determinations) "positive anomalies" cover already a large part of the polygon, the values that seemed to be background values at small scales turn out to be developed in a smaller area. Additional "negative anomalies" are visible. The general view of the pattern of isolines, the extension of anomalies has changed significantly compared to a smaller scale. A new significant change in the pattern is observed when moving to a scale of 1 : 500,000 (400 determinations). The intensity of positive anomalies has sharply increased, their predominant extension and clear confinement to the main geological structures of the polygon were determined.

Moving to a scale of 1 : 200 000 (2500 definitions) allows to highlight numerous details of the field (and to introduce additional TP isolines drawn through 5 mW/m2), but no longer changes its regional structure. More detailed maps can be constructed at individual sites. One of them is for a 1 : 200 000 map cut in the area of the maximum anomaly

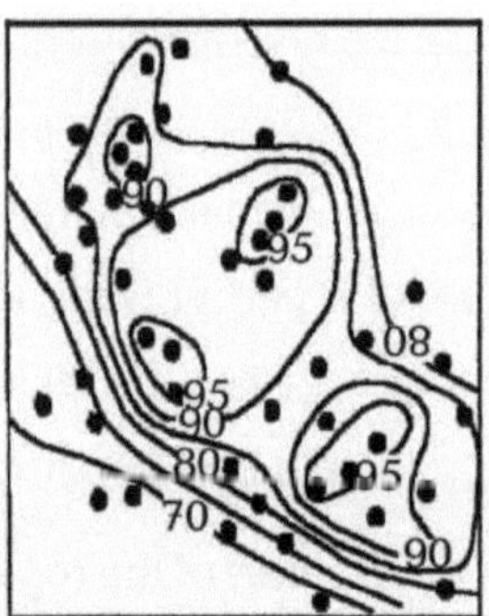

Fig. 2.3. Depth TP map of the north-western end of the Main anticline of Donbass (8x8 km area). shown in Fig. 2.3. The detail of the study corresponds to the scale of 1 : 50 000. Obviously, the regional component (anomaly of northwest strike) is preserved. Details indicating the existence of anomaly sources perpendicular to the main (northeastern) strike are visible.

Thus, stable regional features of the field of real complexity are detected at a

scale of 1 : 500 000 (network density - 4 definitions per 100 km^2 , i.e. 2 orders of magnitude higher than the "average European"). The average density on the territory of Ukraine is 2 times lower, taking into account the concentration of half of the points in Donbass, we can talk about the reliability of detection of the main (regional) field structure approximately only on 30-40% of the area.

The maps show the division of Ukraine into two large regions with markedly different background TP values. These are the most part of the USH, the LDV, the slope of the Voronezh massif and some other regions with the average TP of about 45 mW/m^2 and the territory of the Volyno-Podolsk plate, the southern slope of the shield, the Azov massif, the South Ukrainian monocline, part of the Scythian plate and the slope of the Voronezh massif north of Donbass with the average depth TP slightly more than 50 mW/m^2 . Against this background there are anomalies of intensity 10-45 mW/m^2 of the Precarpathian trough, Volyno-Podolsk plate, Scythian plate, Donbass, UCH, identified with different degrees of certainty. In the extreme west there is the area of high TP of the Carpathian region, where the growth of the parameter from the boundary of the Precarpathian Trough with the Folded Carpathians to the boundary of the Transcarpathian Trough and the Pannonian Depression from about 55 to 85 mW/m^2 is observed. And on this "background" positive anomalies of intensity 10-45 mW/m are also recorded .[2]

The only large negative TP negative perturbation was identified in northwestern Ukraine, near the border of the Volyn-Podolsk plate and the slope of the USH. The anomaly continues in the territory of Belarus.

2.2. Ukrainian Shield, Azov Massif and their slopes

In total, about 1800 single TP values have been established within the region. They are grouped in 350 points. Such a high degree of averaging is mainly explained by the fact that about 1500 single values were determined on a relatively small territory (area of about 10-12 thousand km^2) in the central part of the shield in shallow boreholes at local sites. The average density of the network in the region, which occupies about 40% of the territory of Ukraine, is small. Given the concentration of the main amount of data in the center, it is possible to consider most of the shield unexplored.

Therefore, only a preliminary assessment of the thermal field of the region as a whole is possible. The histogram of TP distribution in the SC reveals the mixing of at least two data sets. An array of background values (about 75% of all determinations) with a modal value of 45 and a standard deviation of 6 mW/m^2 is clearly distinguished. The array of elevated values shows a distribution that can be approximated with some probability by a normal distribution with a modal value of 61 and a standard deviation of 7 mW/m .^{2}

The TP isolines on the map (Fig. 2.4) were drawn at 10 mW/m^2 , which is close to twice the standard deviation for the main data set. The lack of exploration has led to the need, in some cases, to highlight with separate isolines areas where only two points with TPs different from the surrounding ones are encountered. This emphasizes some features of the field, but at the same time introduces a certain degree of subjectivity. The same can be said about delineating with isolines a group of points at a considerable distance from which there are no other TP definitions.

Reliable negative anomalies are not distinguished within the shield. Only in the extreme northwest is shown a part of such a disturbance located mainly on the Volyn-Podolsk plate.

The map of the shield depth heat flux shows that reliably diagnosed positive anomalies (except for several high TPs in points clearly gravitating towards Donbass, where TPs are increased compared to the shield over a large area - see below) are concentrated in the zone of the Kirovograd deep fault and intersecting faults of northeastern strike (Kirovograd TP anomaly).

The Dnieper anomaly stretches on the northeastern frame, in some parts of which the TP exceeds 60 mW/m .^{2}

Some relative increase in TP is observed at the

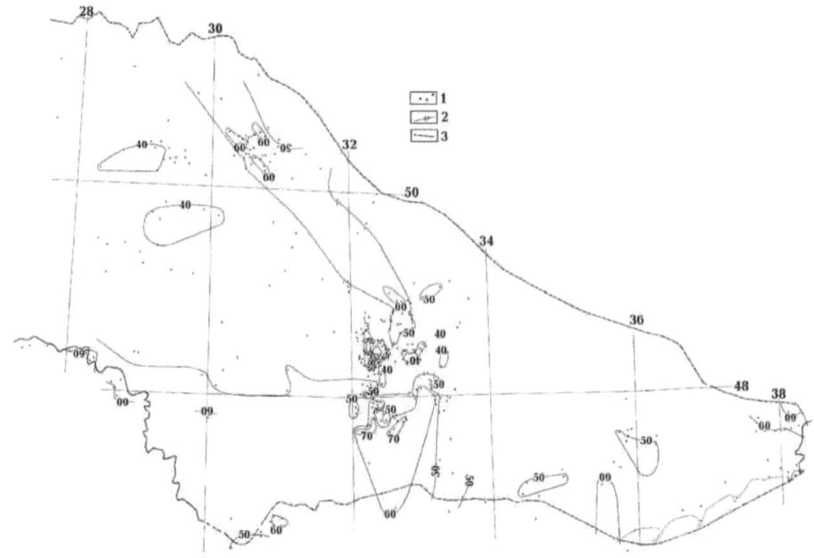

Fig. 2.4. Map of the deep heat flow of the Ukrainian Shield.

1 - measurement points, 2 - TP isolines, 3 - region boundaries.

periphery of the shield, where the positive anomalies of the South Ukrainian monocline and the southern part of the Volyno-Podolsk plate approach its slopes. Values at the level of 50-52 mW/m^2 are common in this area. In some points, higher TPs (up to 65-75 mW/m^2) are found, possibly indicating the existence of positive anomalies not yet studied here.

2.3. Dnieper-Donets Depression

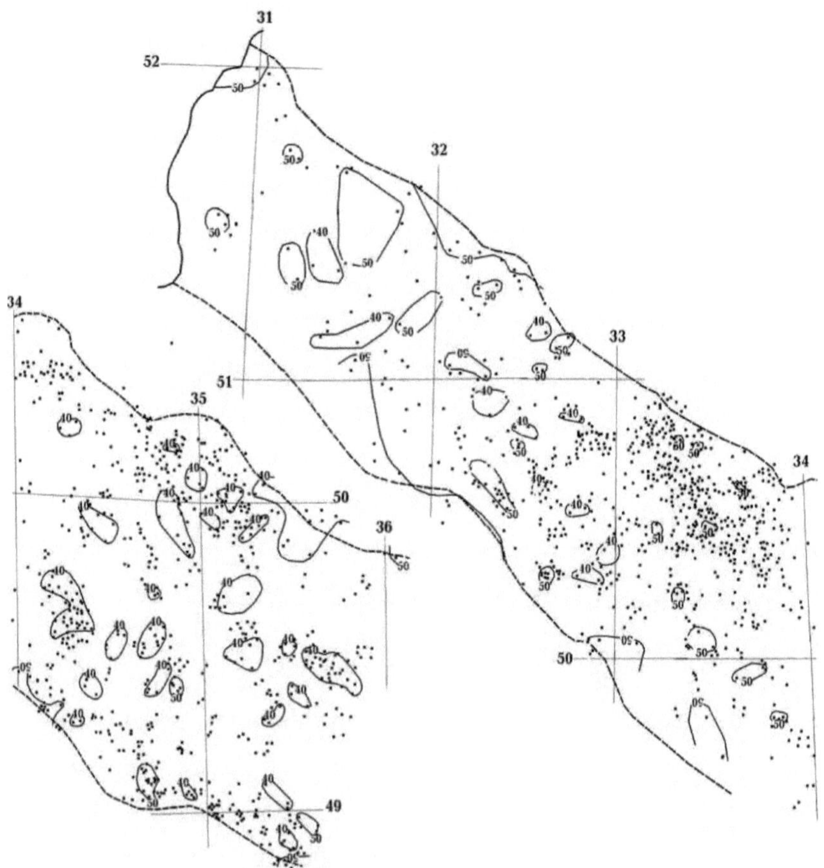

Fig. 2.5. Map of the deep heat flow of the Dnieper-Donets Basin.

See Fig. 2.4 for reference designations.

In the Dnieper-Donets Basin, more than 2,500 TPs have been determined, mostly in the last 20-30 years. Surface T according to meteorological data increases from northwest to southeast from 6 to 9.5^0 C. The average thermal conductivity is obtained as 1.65-1.67 W/m^0 C when the boreholes intersect only Meso-Cenozoic sediments. If Carboniferous and Permian sediments are present in the section, - 1.67-1.95 W/m^0 C. When wells reach Devonian effusive-sedimentary rocks containing salt interlayers, it increases to 1.95-2.05

W/m^0 C on average. When part of the borehole section includes significant layers of Devonian salt, the average thermal conductivity can reach 2.2 W/m^0 C.

To eliminate near-surface distortions of the deep TP, corrections were introduced to account for the influence of paleoclimate and groundwater overflows. The determination error was established by comparing the calculated average values with the values given in [18, 29, etc.], where many results were obtained from mature wells in which high-precision temperature measurements were made along the entire borehole, core samples were taken, and thermal conductivity was determined. The results of comparison indicate an error of the used method of about 7%.

In the north-western part (Desnyansky basin of the DDS), the histogram of the TP distribution of the region allows us to outline two data sets. The first one (the larger one - about 85% of all TP values) is characterized by a modal value and standard deviation of 45±5 mW/m^2 . It obviously describes the background. The second one - 55±5 mW/m^2 refers to positive anomalies. The increase in TP is observed along the northern and southern sides. A chain of "negative anomalies" of TP (less than 40 mW/m^2) stretches along the central part of the area. The confinement of the TP anomalies of both signs to the fault zones of the crystalline basement is characteristic. It can be assumed that above the faults there are permeable zones in the cover rocks, through which groundwater rises (in positive TP anomalies) or descends (in "negative anomalies"). Thus, we admit that the hydrodynamic scheme of the region near the faults used to introduce the hydrogeological correction does not accurately describe the real situation. Without denying this possibility in principle, we mention that the "negative anomalies" (with an average TP value of 37 mW/m^2) differ from the background by less than twice the standard deviation, and cannot be considered as reliably identified heat flux perturbations of the region.

The part of the DDP located to the southeast of the considered part and the transition zone from the DDP to Donbass are united under the name of the "Dneprovsky oil and gas bearing basin". Within its boundaries, the thickness of the Paleozoic sedimentary cover grows to the southeast, in the same direction the role of solanic-dome tectonics increases, in the southeastern part the influence of the Donbass structural plan is noticeable.

The histogram of the TP distribution of the northwest basin demonstrates its character close to normal for the main part of the data set with a modal value of 44 mW/m^2 and a standard deviation of 2.7 mW/m^2 . Obviously, there are no reliable negative anomalies in the region, while positive anomalies (TP inside the 50 mW/m^2 isoline averages 57 mW/m^2 , at maximum - 65-67 mW/m^2) are localized and mostly confined to faults. The influence of faults appears to be different. Probably, the upward movement of deep water is mainly associated with them, leading to positive disturbances of TP, but in some areas infiltration (more intense than that accounted for by the hydrogeological correction) may prevail . In this case, small depressions of TP occur.

In the central part of the Dnieper basin in the area bounded approximately by meridians 33^0 30'-35^0 , the histogram of TP distribution reveals its character close to normal. The modal value is 43 mW/m^2 , the standard deviation is 4 mW/m^2 . Accordingly, reliable negative anomalies are not recorded, positive ones are represented by several small area disturbances with average values (within the isolines of 50 mW/m^2) of about 51-52 mW/m^2 . Nevertheless, at the qualitative level (within the isoline 45 mW/m^2 the average TP value is only 48 mW/m^2 , i.e. exceeds the background by only one standard deviation) we can state the existence of an extended positive anomaly transverse to the Dnieper-Donets depression in the western part of the region. It is located within the block bounded by the West-Inguletskiy and Krivorozhsko-Kremenchug transverse faults. The Kishinev lineament zone stretches in the central part of the block [28]. The southwestern marginal fault of the DDV is not traced within the block, while the northeastern one is sharply bent, probably due to displacements by transverse relatively young faults. The anomaly continues the Kirovograd anomaly established in the Ukrainian Shield (see above).

In the southeastern part of the basin and in the transition zone to Donbas, histograms of TP distributions were plotted for two distinctly different thermal field areas located approximately east and west of the line connecting the points with coordinates 50^0 N, 37 E, and 49 N, 37 E, and 49 N, 37 E, and 49 N, respectively. - 37^0 E and 49^0 N. - 36^0 E.

The modal values and standard deviations for the predominant data sets in the districts are 42±5 mW/m^2 and 50±5 mW/m^2 , respectively. Small positive anomalies in each district are seen, negative anomalies are not reliably recorded.

Comparison of TP values determined by different methods within the region allows us to estimate the error of the main method used (calculation of TP by the average geothermal gradient between bottomhole and surface and effective average thermal conductivity of rocks of this depth interval) to be about 8%. Accordingly, TP isolines were carried out at 10 mW/m^2 , auxiliary ones at 5 mW/m^2 , but the 55 mW/m isoline^2 was not carried out.

As in other parts of the VDV, positive anomalies are largely confined to marginal deep faults (and others, often transverse to the trough strike), with a band of relatively low TP values extending through the central part of the region.

The values of the depth heat flux within the territory of the LDV in general are 43.5-44.5 mW/m^2 (depending on whether or not the increased TPs in the transition zone from the LDV to Donbass are taken into account). It does not noticeably differ from that set in the Ukrainian shield.

2.4. Slope of the Voronezh massif

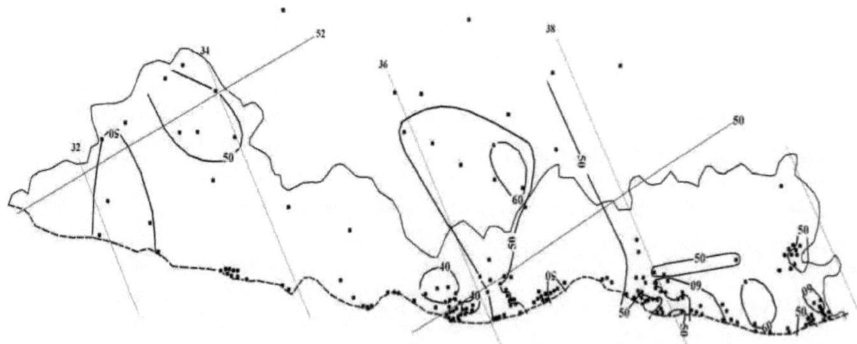

Fig. 2.6. Map of the deep heat flow of the Voronezh massif slope.

See Fig. 2.4 for reference designations.

In the extreme northeast of the territory under consideration, the map also includes a part of the Voronezh massif proper with minimal (up to the first hundred meters) sediment thickness.

When compiling the map (Fig. 2.6) of this relatively poorly studied region, the data within Ukraine were supplemented with data on the heat flux in Russia, mainly obtained by the authors.

The errors of TP determination on the slope of the Voronezh massif are at the level of 3 mW/m^2 . TP isolines can be drawn through 10 mW/m^2 . But at the present stage of research, when "white spots" are still widespread in the territory, it makes sense to construct, first of all, an overview scheme of heat flux distribution, within the framework of which many values in neighboring wells cannot be represented as separate points. As a rule, such data were subjected to averaging. As a result, the scheme shows TP values in 163 points in Ukraine and 13 points in Russia. Naturally, with such a sparse and highly irregular network, heat flux isolines cannot be performed as part of a conditioned mapping procedure. In the presented scheme, they are intended to somehow structure the available data, to contribute to the preliminary allocation of heat field disturbances (although not for the whole territory yet, with probable omissions).

Background distributions are characterized by modal TP values of 43 and 50 mW/m^2 , anomalous ones - 53 and 60 mW/m^2 . The magnitude of the inferred anomalies - 10 mW/m^2 - is quite common among regional disturbances in the zones of modern activation in Ukraine and other regions [16 et al.] However, such a coincidence cannot be considered sufficient justification for the nature of the anomalies; they may also be related (at least near the DDV) to variations in radiogenic heat generation of crustal rocks.

In the first case, the background value does not noticeably differ from the usual one for the Precambrian platform, while in the second case, it is close to the background value in Donbass (although slightly exceeding it). But in Donbass, the background is caused not only by radiogenic heat generation in the crust and normal TP from the mantle, but also (to a small extent) by the influence of the cooling interior of the Hercynian geosyncline. The latter probably does not apply to the slope of the Voronezh massif. Although this option cannot be excluded: according to some authors, Hercynian magmatism is quite widespread on the massif.

According to the available data on the slope can be identified preliminary 6-7 positive anomalies, but their reliable definition and delineation makes sense to carry out after comparing the experimental TP with the calculated ones. Reliable negative anomalies are not established in the region: within the isolines of TP 40 mW/m^2 average values are 37-38 mW/m^2 , i.e., differ from the background by less than double the error of TP determination.

2.5. Donbass

The thermal conductivity of rocks in the region has been studied more fully than in other places. In areas where Carboniferous sediments are not practically overlapped by sediments of Mesozoic and Cenozoic age, the average value of thermal conductivity varies from 1.95 to 2.04 W/m^0 C. The thermal conductivity is somewhat lower (1.65-1.9 W/m^0 C) in the zone transitional to the Dnieper-

Donets Basin (including the Kalmius-Toretskaya and Bakhmutskaya Basins). A sharp increase in the vertical component is observed in the Main and Druzhkovsko-Konstantinovskaya anticlines

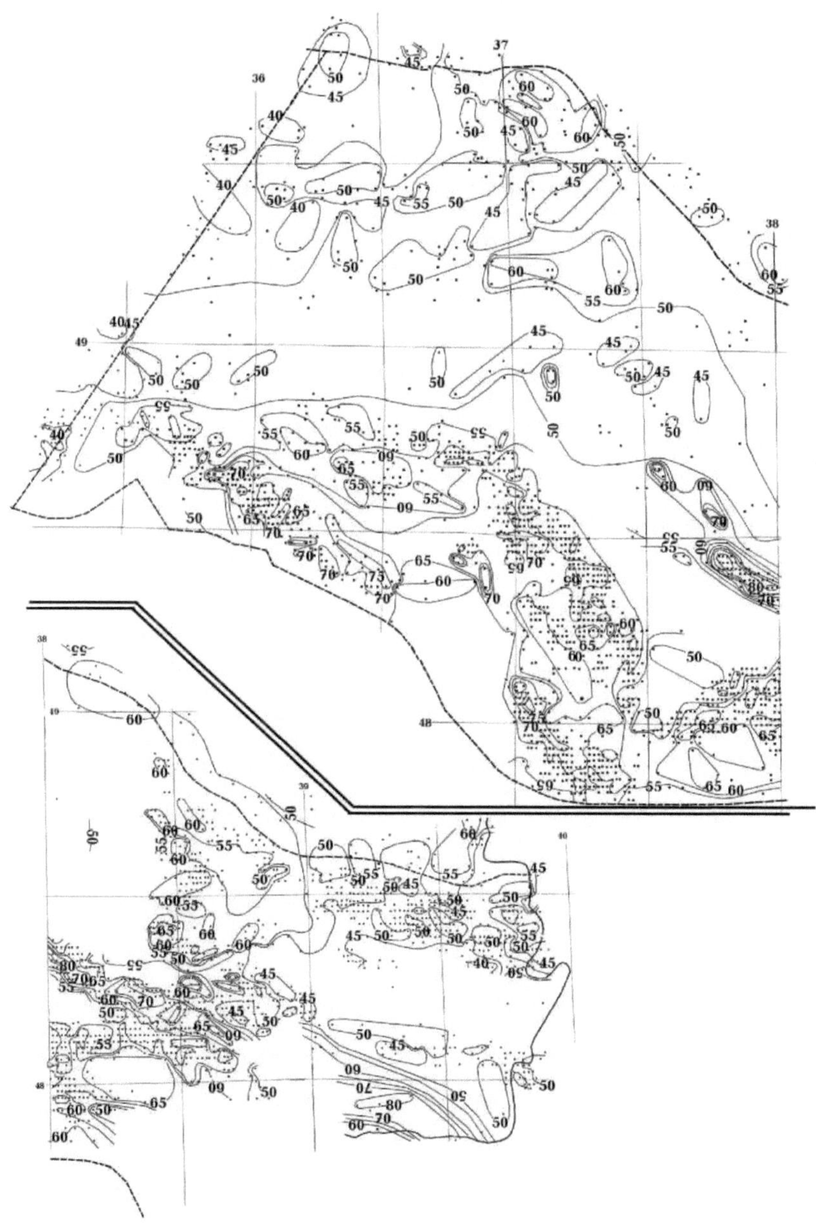

Fig. 2.7. Map of the deep heat flow of Donbass.

See Fig. 2.4. (up to 2.3-2.6 W/m^0 C) due to the anisotropy of thermal conductivity of the layered strata, as well as an increase in the proportion of sandstones in the section.

The values of downhole temperatures were corrected to account for paleoclimate changes. At the Main and Druzhkovsko-Konstantinovsko anticlines it was necessary to introduce a structural correction, which is -7 mW/m^2 at the fold lock, -4 mW/m^2 at the wings and smoothly increases to +1 mW/m^2 at the transition to the troughs. No hydrogeologic correction was introduced because it has been previously established that there is no appreciable influence of downwelling water in the region at the depths of T measurements.

Comparing the results of repeated TP determinations by different methods results in an error of 2-3 mW/m^2 . Such accuracy allows to draw isolines through 5 mW/m .^{2}

The results of TP determinations averaged within the territories bounded by 1'x1' coordinates are presented in Fig. 2.7. Several TP determinations in the Russian part of Donbass made by the authors are presented. They provide an opportunity to more confidently draw isolines of the deep heat flux near the border of the states. A total of 2750 points of TP determinations were used. The map allowed to reveal many patterns of TP anomalies location on the slope of the Ukrainian Shield, the southern edge of Donbass, and the Main anticline, which were not noted before.

The largest part of the territory is characterized by heat flux values from 42 to 52 mW/m^2 (Kalmius-Toretskaya and Bakhmutskaya Basins, the territory north of the Nagolny Ridge to the Northern zone of shallow folding). In the Northern zone of shallow folding the heat fluxes increase to 48-55, rising to 62 mW/m^2 at the junction with the Bakhmut Basin, on the slope of the Voronezh Massif they are 52-56 mW/m^2 . TPs are maximal on the slope of the USH, the Main and Druzhkov-Konstantinovskaya anticlines, in the Donetsk-Makeevsky area. Here the average values are 63, and the maximums reach 75 mW/m^2 in

Donetsk-Makeevsky area, 103 mW/m^2 - on the slope of the shield, and 104 mW/m^2 at the Nikitovskoye ore field of the Main anticline.

The background values of 48-49 mW/m were taken as the most common values over the area^2 . The obtained value slightly exceeds the calculated and observed background heat flux on the Ukrainian Shield and the Voronezh Massif and is close to the value established in the transition zone between the DDV and Donbass. In almost all cases, the anomalies extend along the main fold structures of the basin and have a significant length (up to 140 km) with an insignificant width (from 4-10 to 20 km), separating areas with relatively low TP values. Narrow anomalies correspond to longitudinal deep faults, and background values are characteristic of blocks not disturbed by disjunctives. The dependence of the anomalies on the location of fault zones (rather than plicate structures) is clearly revealed in the southwestern part of Donbass.

Fault zones and their intersections are manifested not only in the increase of modern TP associated with the removal of deep heat and fluids (including hydrocarbons), but also in magmatism, hydrothermal ore mineralization (gold, mercury, etc.) of past activations. When they are crossed, the degree of lithification of sedimentary rocks (including coals) changes abruptly due to changes in the level of erosional shearing on different blocks of the basement. Therefore, identification of fault zones is an important task in mineral prospecting. It can be solved by means of detailed TP maps. In the areas covered by the modern depth process, they will be manifested by TP bursts (experience shows that they are confined to only a part of the length of the activated fault). In the absence of modern activation and/or inflow of deep heat (e.g., in the place where the upward branch of the convective cell approaches the surface), fault zones are manifested by insignificant (up to 5 mW/m^2) negative TP anomalies .

2.6. South Ukrainian monocline and Scythian plate

This region comprises two tectonic units different in geologic history - the platform South Ukrainian monocline (occupying, apart from the land part in the south of Ukraine, also a part of the shelf), the Priborudzhsky trough and fragments of the Hercynian-Cimmerian geosyncline of the Scythian plate on land (part of Northern Dobrudja and Crimea) and on the shelf.

The average thermal conductivity of rocks was established by values for the main lithologic-stratigraphic horizons and varied in the range from 1.45 to 1.95 W/m$^{\cdot 0}$ C. The error of TP determination was established by comparing the obtained values with those established earlier using other techniques. The discrepancies amounted in most cases to 10-15% of the average value, indicating an error of each of the determinations of about 9%.

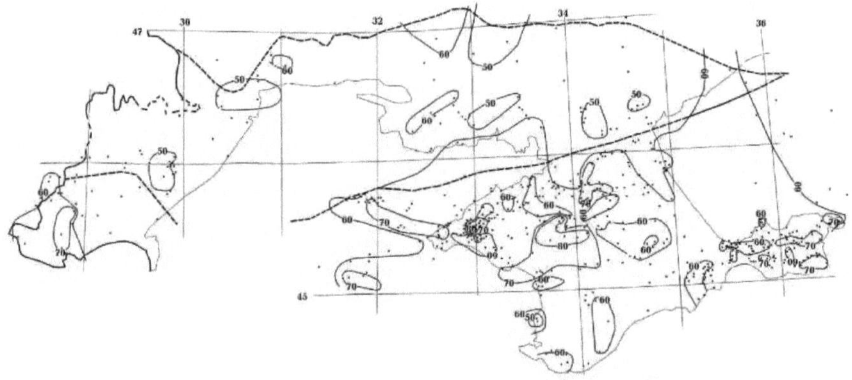

Fig. 2.8. Map of the deep heat flow of the South Ukrainian monocline and the western part of the Scythian plate

See Fig. 2.4 for reference designations.

On the monocline, the deep TP was determined in 300 boreholes grouped in 200 points. The heat flux data in the Crimean points closest to the shelf were used to draw the TP isolines on the map. Isolines of heat flux on the map were drawn at 10 mW/m^2 , which in some cases is comparable to the doubled error, but in areas with high values is less than it. Therefore, we can talk about the qualitative nature of isolation of the most intense parts of positive TP anomalies (where values exceed 70 mW/m).^{2}

For most of the study area - the South Ukrainian monocline - the background value of TP is about 50 mW/m^2 . In the extreme west of the region stands out Renia positive anomaly, in the center of which the heat flux exceeds 70 mW/m^2 To the east is located Shelf anomaly of approximately the same intensity. To the north of the Crimea, the relative increase in TP (more than 60 mW/m^2) extends towards the southern end of the Kirovograd anomaly identified on the shield. Noticeable TP increases are found on the northern margins of the Crimea (see below).

In general, the increase of the deep heat flux with approach to the Scythian plate, complicated by intense local positive perturbations, is noticeable.

On the slab, the thermal conductivity of rocks in the depth intervals of the Crimean TP calculation was calculated using numerous data obtained with the participation of the authors. The available information allowed us to present the change in the value of thermal conductivity with depth for typical cases as: up to 1.5 km - 1.6, up to 3 km - 2.05, up to 4.5 - 2.5, deeper - 2.65 $W/m \cdot {}^0 C$. However, exceptions to this dependence are quite widespread, especially in the areas of the Evpatoria and Novoselovsky uplifts, where thermal conductivity is higher in the upper part of the section. On the contrary, in the Indolo-Kuban trough, the sole of the upper stage, composed of rocks of low thermal conductivity, is lowered to a greater depth. The observed TP values were corrected for the influence of paleoclimate. TP distortions related to the structural effect and groundwater movements were taken into account only in a small number of points; for the rest of the territory they were recognized (at the real TP calculation depths) as insignificant.

The total number of single TP determinations amounted to 600, they are grouped in 370 points. Comparison of TP values determined by different methods indicates a probable error of about 10%.

The histogram of the Crimean GTP values suggests, quite reasonably, that it reflects the presence of two data sets. The first, comprising about 75-80% of all values, probably represents the background distribution, characterized by a modal value of 60 mW/m^2 and a standard deviation of about 6 mW/m^2 . The remaining TPs belong to positive anomalies and their distribution can be roughly characterized by a modal value of 73 mW/m^2 and a standard deviation of 8 mW/m^2 . Obviously, negative anomalies (TPs less than 48 mW/m^2) are not reliably distinguished. Several groups of positive anomalies (which were considered to be areas of development of TP values more than 70 mW/m^2) were detected. These are the Tarkhankut group in the west of the peninsula (82±6 mW/m^2), Novoselovsko-Evpatoria (78±7 mW/m^2) - in the center, Kerch (74±5 mW/m^2) - in the east. To the north of the central group, two more relatively small in area TP disturbances can be distinguished, forming the Sivash group (74±5 mW/m).^{2}

Comparison of positive anomalies and relative TP depressions with the grid of the main faults of the peninsula suggests that here (as in the Dnieper-Donets Basin and Donbass) the thermal field perturbations gravitate to disturbances. Although the upward and downward transport of heat together with fluids can hardly be recognized as the only cause of disturbances. Some of the positive anomalies of the Tarkhankutskaya and Novoselovsko-Evpatoria groups are located at a noticeable distance from the faults. Apparently, in these cases, the conductive heat transport also plays a significant role. The confinement of other anomalies to faults seems more obvious.

2.7. Volyno-Podolsk plate

The region includes, besides Volyn-Podillya proper, a fragment of the West European plate with the Caledonian-Hercynian basement in the basement of the Lviv Paleozoic trough.

The average value of thermal conductivity was assumed to increase with depth from 2.1 W/m$\cdot^0$ C at bottomhole depth of about 700m to 2.4 W/m$\cdot^0$ C at bottomhole depth of 4000m and more. The TP was calculated from these parameters. The accuracy of this operation was established by comparing TP for several reference wells where TP determinations were made by different methods. The results of the comparison indicate an error of each method of about 10%.

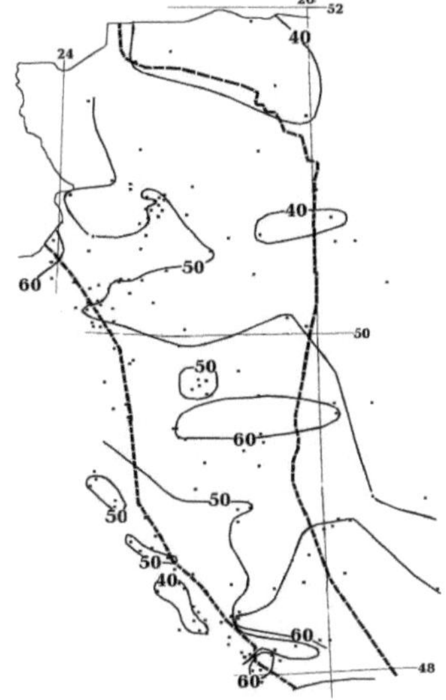

Fig. 2.9. Map of the deep TP of the Volyn-Podolsk plate.

See Fig. 2.4 for reference designations.

No appreciable structural effect has been established in the region under consideration, and the corresponding correction has not been introduced. There is little information on groundwater overflows at depths greater than the first tens of meters; the data known to the authors indicate a small amount of distortion of the thermal field by water. Although it cannot be stated that this was correct in all cases, a local increase in the error of TP determination is not excluded. Hydrogeological corrections were introduced into the results of studies in shallow boreholes and noticeably affected the value of TP (they reach many tens of percent of the observed value). Paleoclimatic correction was introduced everywhere: from 2 to 13 mW/m^2 depending on the depth of the well.

In total, about 250 TP values were established, grouped in 160 points. The map of the deep TP of the region allows us to outline such a picture of its distribution (Fig. 2.9): background values, three positive anomalies (two of them in fragments) - Yavorivska in the northwest of the Lviv Trough, Chernovtsy in the south of the Ukrainian part of the plate and Ternopil in the center of the plate - and one negative anomaly - Volynska in the north of the plate, which has a continuation in the territory of Belarus - are presented here.

Outside the anomalies, the TP averages about 46-47 mW/m^2 , which is close to the background values in other platform regions of Ukraine (see above). However, two areas of slightly different moderate heat fluxes are clearly distinguished in the study area. One of them is directly adjacent to the slope of the Ukrainian Shield and is characterized by a TP of about 43 mW/m^2 . The second one is located to the west and here the average heat flux reaches about 50-51 mW/m .^{2}

It is on the "elevated background" that the positive anomalies are developed. The direct influence of heat sources of the Carpathian anomalies cannot explain the general increase of TP in the western part of the region. This is proved by appropriate calculations. Probably, the thermal effect of modern activation, with which the positive TP anomalies are clearly associated, in the rest of the western part of the region is only beginning to manifest itself. In the areas of these positive anomalies (average TP - about 63 mW/m^2), the manifestation of thermal effects of modern activation is undoubted. Studies of the TP in Poland and Moldova, conducted by the authors, established that (perhaps, after some break) the Jaworowski anomaly continues in the northeast by the Chelmskaya TP anomaly on the territory of Poland, and the Czernowitz anomaly in the southeast by the Balti anomaly on the territory of Moldova.

2.8. Carpathian region

Three major structures can be distinguished within the region, differing in the history of geological development and thermal field: 1) Pre-Carpathian Trough, which emerged at the last stage of active processes in the Carpathian-Dinaridian alpine geosyncline, superimposed partly on the edge of the folded zone (Inner Trough Zone), partly on the platform foreland of the geosyncline (Outer Trough Zone), 2) Folded Carpathians - typical alpid externides, 3) Transcarpathian Rear Trough, formed on the margin of the median massif during the last period of the geosyncline development (Fig. 2.1).

In the Pre-Carpathian trough, 500 TP values were established, grouped in 220 points. The values of thermal conductivity increase markedly with the depth of the bottom of the calculated interval, which made it possible to construct a rather simple dependence of the average value on this depth. Thermal conductivity increased almost linearly from 1.7 $W/m \cdot {}^{0}C$ for 300-500 m to 2.2 $W/m \cdot {}^{0}C$ for 5000 and more meters. Only in rare cases of inclusion of thick salt layers in the section did the average thermal conductivity increase to 2.5 $W/m \cdot {}^{0}C$.

Paleoclimatic corrections were made to all values of depth T. Corrections taking into account groundwater overflows were not introduced due to the lack of necessary information. Naturally, the errors of TP determination could be related to this fact: underestimation of the parameter in relatively shallow wells. However, in those numerous cases when several TP values were determined in one well at different bottomhole depths, there was an obvious dependence of TP

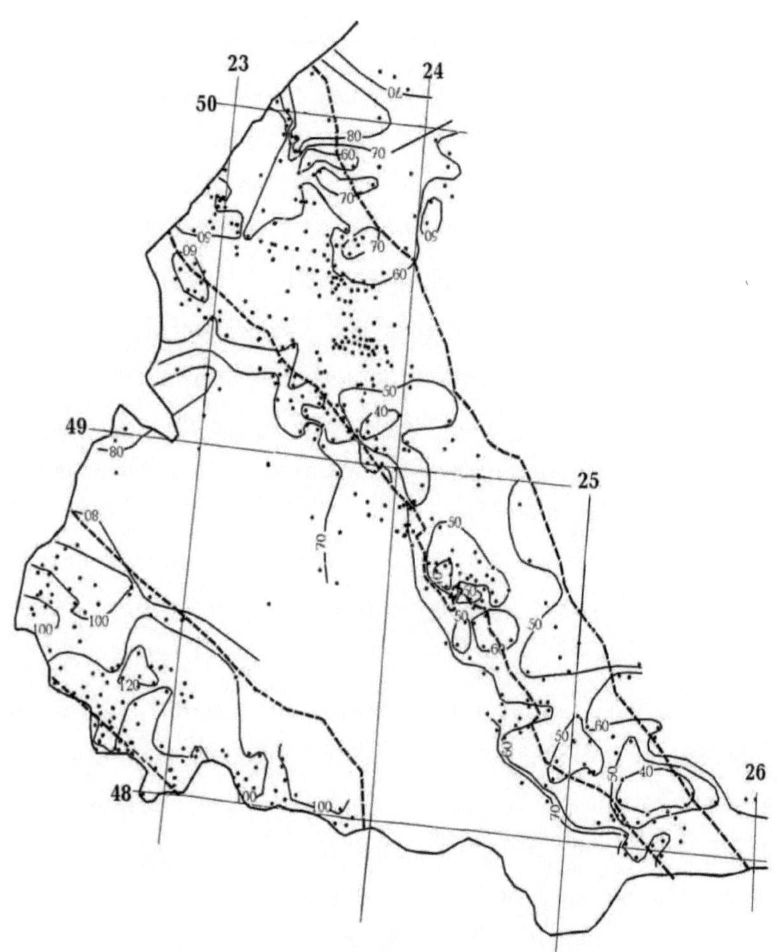

Fig. 2.10. Map of the deep heat flux of the Carpathians See Fig. 2.4

from depth has not been detected. The structural correction due to the fact that the thermal conductivity of the molasse filling the sag is less than the thermal conductivity of the underlying rocks (flysch formations of the Folded Carpathians and ancient rocks of the Volyno-Podolsk plate) for most of the sag is (according to the results of estimation calculations) vanishingly small - about 3-4%. The only exceptions are the parts of the trough directly adjacent to the faults bounding it. However, there are almost no TP definitions for the Precarpathian fault. Thus, it can be assumed that in the course of the studies, a deep (not significantly distorted by near-surface influences) value of TP was established.

The accuracy of the obtained values was determined by comparing the TPs established using different methods. The detected differences in TP values indicate an error of each method of about 8-10%. The absolute error is about 4 mW/m^2 , which allows to draw isolines on the TP map of the region through 10 mW/m .^{2}

The histogram of the distribution of heat flux values shows that three data sets including approximately comparable number of values are quite clearly isolated within the sag. The first is characterized by a modal value and standard deviation of 41 and 2 mW/m^2 , respectively. The second array consists of slightly higher TPs of 49±3 mW/m^2 . The third consists of high heat fluxes - 58± 5 mW/m^2 . Obviously, the first two arrays constitute the background TPs of the Precarpathian Trough. The lowest values are obtained mainly in the 10-km-wide band near the Folded Carpathians thrust, where, under the influence of structural effects, the TP may be underestimated by 10% on average. The introduction of an appropriate correction brings the modal values of the arrays closer together and allows us to estimate the average background TP of the sag at 47 mW/m^2 . The obtained background value is close to that established within the Ukrainian Shield, Dneprovsko-Donets depression and Volyno-Podolsk plate. Therefore, it can be stated that in the studied region the effect of the Alpine geosyncline of the Carpathians is very weak, not exceeding the first mW/m .^{2}

However, a part of the increased TP (included in the third data set) is established in a 10-km strip near the thrust of the Folded Carpathians and can be related to the increase in heat flux when approaching the Folded Carpathians, at the edge of which TP is estimated at 55-60 mW/m^2 . In this zone, after the above correction, the average TP reaches 53 mW/m^2 . Therefore, it is logical to assume that the TP across the sag still varies, increasing from the northeastern border to the southwestern one by 5-10 mW/m .^{2}

Geothermal studies in the Folded Carpathians are insignificant compared to the Pre-Carpathian and Transcarpathian troughs. This is due to the small number of wells in the inner part of the folded zone.

For the Cretaceous-Paleogene flysch, a thermal conductivity value of 2.65 W/m$\cdot^0$ C was assumed, for rocks of the Precarpathian Trough and some other formations of the western part of the Folded Carpathians - 1.92.0 W/m$\cdot^0$ C. In the thrust zone of the Skibovy Carpathians on the Predkarpattya Trough, the average values of λ in the interval of geothermal gradient determination were usually about 2.3 W/m$\cdot^0$ C.

Paleoclimatic corrections were made, which noticeably affected the TP values in shallow boreholes and practically did not change them in deep boreholes (the vast majority of boreholes 2000-4000m deep).

The accuracy of the obtained TP values can only be roughly estimated by the scatter of values in one well, in neighboring wells, by the results of comparison with TP established by another methodology in wells where repeated determinations were carried out. Deviations from the average in the latter case are 6-7%, which indicates a small level of error and with the average value of background TP of the region (see below) of about 65 mW/m^2 allows to draw TP isolines on the map after 10 mW/m^2 .

120 points (300 single TA definitions) were mapped.

It seems logical (until the appearance of data in the center of the southeastern half of the Folded Carpathians) to consider the increase in TP from the Precarpathian Trough to the Transcarpathian Trough as less intensive - by about 10 mW/m^2 . It is also necessary to take into account that in the extreme zones near the Pre-Carpathian Trough the heat flux is probably overestimated by the structural effect by about 3-4 mW/m^2 , i.e. the increase in TP across the

Folded Carpathians is about 15 mW/m^2 (from 55-60 to 70-75 mW/m^2). And against this background, positive anomalies are developed, one of which is located almost entirely in the Transcarpathian Trough, the second - on the northwestern margin of the region. Within its limits, the average TP reaches 82 mW/m^2 , in the central part - more than 90 mW/m .^{2}

The study of TPs in the Folded Carpathians outside Ukraine is very insignificant. Nevertheless, it can be stated that the distribution of TP in the Western Carpathians in Poland is generally similar to that observed in Ukraine. On the contrary, in the Romanian Eastern Carpathians the TP of the folded zone is sharply lowered compared to the considered one.

Average values of thermal conductivity of rocks of the Transcarpathian sagging were determined from the borehole section. As a result, it turned out that the parameter varies from 1.6 to 2.3 W/m$\cdot^0$ C. The exception were a few boreholes that crossed salt strata significant in thickness. Here the average thermal conductivity reached 2.7-2.8 W/m$\cdot^0$ C.

At several points, the TP determinations were repeated in previously studied wells, with deviations from the average value of 5-20%. The calculated TP values were corrected to account for the influence of paleoclimate, and at several points for relatively shallow wells the influence of water overflows, especially intense in the upper part of the section, was taken into account. As a result, there is no dependence on the depth of temperature determination in the obtained TP values (in some wells there are up to 5-7 of them), which indicates the efficiency of the corrections calculation. It can be considered that a deep (i.e., not significantly distorted by surface influences) heat flow is established in the region. TP isolines are drawn at 20 mW/m .^{2}

The area of regions where the heat flux exceeds 100 mW/m^2 , is about 1/3 of the territory. Outside the corresponding isoline, the average value of TP is 89 mW/m^2 . This value can be recognized as a background value for the Transcarpathian Trough. According to a small amount of data, it can be assumed that it is characteristic for the fragment of the Pannonian Trough on the territory of Ukraine. The named value is noticeably larger than that observed in the part of the Folded Carpathians adjacent to the trough (78 mW/m^2). Thus, the TP growth within the Folded Carpathians in approaching the Transcarpathian Trough sharply increases and is most likely related not to the "all-Carpathian", but to the local source.

This regional increase in TP in the sag itself is complemented by intense positive anomalies. Within their limits, the average value of heat flux is 113 mW/m^2 . In small areas (about 3% of the sag area) the TP exceeds 120 mW/m^2 (on average - about 130 mW/m^2). Obviously, in these areas the heat sources are sharply close to the Earth's surface.

The above data show that the heat flow of the Transcarpathian Trough can be considered as a result of the action of heat sources additional to a single source related to the deep processes in the interior of the Alpine Carpathian-Dinaridian geosyncline. One (regional) source causes a step-like increase in TP at the outer boundary of the sag. Here the heat flux increases by 15-20 mW/m^2 within a band about 15 km wide. Others (local) form anomalies with a characteristic width of about 15 km, in the central parts of which the TP is increased by 20-40 mW/m^2 compared to the one caused by the regional source. In a less detailed and reliable form, the same picture is recorded in the Pannonian Median Massif (young Pannonian Depression).

Chapter 3. Thermal models of the crust and upper mantle

3.1 Initial data and principles of model building The focus of the work predetermines the depth interval for which the models are built. It is limited from below by the central part of the subcrustal layer in the mantle, which is overheated in active regions and can manifest itself in anomalous temperatures and geothermal resources near the surface. Thermal models (i.e., distributions of heat sources and temperatures) are created as a result of interpreting the distribution of TP across the Earth's surface. Interpretation involves solving an inverse problem, naturally ambiguous. For geothermia, the ambiguity is especially great due to the fact that a heat source in a certain period may not be sufficiently manifested or not manifested at all in the studied heat flux. Therefore, it is necessary to use specific methods of regularization of the solution of the inverse problem, involving non-geothermal data. The authors have developed an interpretation scheme including such stages.

1. As a means of regularizing the solution of the inverse problem, a model of the crustal and upper mantle matter compositions and deep processes in them is used (the scheme of the latter corresponds to the advection-polymorphic hypothesis used by the authors [8, 11, etc.]). The model is, of course, hypothetical, but modern geological and geophysical data allow its thorough verification.

2. The model of heat sources (stationary and non-stationary) is built by the model of compositions and processes. It is used to solve the direct problem. The result is compared with the TP through the surface. This should achieve a match within the limits due to the error of the experimental material and the calculation. The calculation error is considered to be related only to the tolerances of the model parameters and not to the main features of the model. The selection of the calculation effect for better agreement with experimental data is allowed only with the help of a limited change in the value of 1-2 minor (for the model as a whole) parameters in the models of young nonstationary

sources. This is due to the incomplete manifestation of the modern incomplete deep process in geologic phenomena.

The reliability of the obtained result is determined at two stages of control.

3. First, the thermal model is compared with the data of geologic thermometers. In addition to the traditionally used information on xenoliths, other data can be applied, in particular, on the depth of magmatic centers and the temperature of melt in them, on the distribution of temperatures by mineral paragenesis of hydrothermal formations, and others.

4. The final control is based on geophysical data. The thermal model is used to calculate the distribution of physical properties of crustal and mantle matter, or more precisely, their anomalousness, i.e., their difference from the properties corresponding to normal temperatures. The results are compared with seismic, geoelectric, magnetic, gravity models directly or (in the latter two cases) through comparison of calculated effects.

It is technically convenient to conduct thermal field interpretation by first developing a normal model and then analyzing anomalous models as a result of changes in the normal model.

In principle, all sources in the real Earth are nonstationary, i.e., their effects change in time. But the degree of changes is different and for certain sources and classes of problems nonstationarity is traditionally and reasonably not taken into account. However, this approach cannot be automatically extended to any time intervals of action and depths of location of such sources. The calculation shows that the thermal field up to a depth of about 70-75 km can be considered practically stationary if the last unsteady sources in the region were active in the Precambrian.

Thermal properties of the medium, which were used for model building, are discussed above (see Chapter 1). The following were also used: heat of melting of mantle rocks - $1.28 \cdot 10^9$ J/m^3 (according to some data it can increase twice in the lower part of the tectonosphere), volumetric heat capacity of rocks and melt - 4.26-10^6 J/m .^{30} C, adiabatic change of T at vertical displacement of matter - 0.5^0 C/km. At real degrees of melting in large volumes of rocks (up to 5%) cooling due to heat of fusion is the first tens of degrees, this temperature change can be neglected in calculations. The heat of polymorphic transformation at the base of the upper mantle is more significant. At transformation of all olivine (about 50% of the rock volume) into a mineral with spinel structure there will be heating by 100^0 C. Completion of polymorphic transitions (to a depth of about 650 km) will result in the release of heat sufficient to heat the rock by another 200^0 C.

The values of radiogenic heat generation in crustal and mantle rocks of Ukraine were also used.

Sedimentary layer. Quite detailed studies of uranium and thorium contents in sedimentary cover rocks have been carried out on the territory of Ukraine. Reference data on potassium contents were also used in the calculations. They did not significantly affect the results of TG calculation. The average density of the studied rocks was about 2.3 g/cm .^{3}

According to such data for about 600 samples in DDV, the average TG of about 1.1-1.2 µW/m was obtained^3 . In Donbass (300 samples), the average is about 1 µW/m).^{3}

In the Pre-Carpathian Trough (approximately 1500-2000 samples), TG values of about 1.0-1.1 µW/m^3 are mostly common, but clays with TG values up to 1.7 µW/m^3 are occasionally found.

On the southern slope of the UZH, the average value of heat generation based on data from about 100 determinations is about 1.3 µW/m^3 . On the northern slope - 1.1-1.2 µW/m^3 . Similar values (about 200 samples) were obtained for the sedimentary cover of the Scythian plate. In the Black Sea muds (when reaching a comparable density to that used in the calculations) the TG is (300 samples) about 1.2 µW/m .^{3}

Thus, heat generation in weakly lithified rocks of the upper part of the sedimentary layer is rather stable. Toward the lower part of the thick layer, where lithification increases significantly, the TG decreases according to available estimates to about 0.8 μW/m^3 , in the limit (at the lithification temperature of 400^0 C) - to 0.6 μW/m^3 [9, 10 and others]. Accordingly, the average heat generation of the sedimentary layer should be determined individually, decreasing, other things being equal, with increasing thickness of the thickness. It turns out to be minimal in the Donbass - about 0.8-0.7 μW/m^3.

Consolidated crust. In the upper part of the granite layer proper (i.e., in rocks with a degree of metamorphism not lower than greenschist), the heat generation is somewhat higher than in the sedimentary layer. The average value turns out to be at the level of 1.5 μW/m^3 , the variability of the parameter is very large. The sizes of fields of relatively stable TG values are clearly related to the sizes of rock massifs on the erosional slice of the shield, i.e., they reach several tens to a hundred km for large plutons, but are mostly limited to several to ten kilometers.

The granitoid massifs associated with high TG values most likely do not cover a significant part of the crustal thickness. The situation may be different in the zones of low TG associated with the development fields of metamorphic rocks of relatively lower acidity. Here, the average value of TG is reduced to 0.8 μW/m^3 . If such negative anomalies persist at least at half of the crustal thickness (naturally, decreasing in amplitude), we can expect a decrease in the crustal component of TH by 10 mW/m^2.

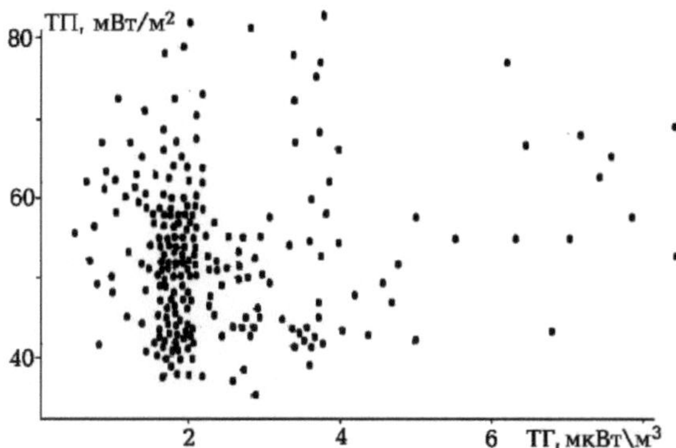

Figure 3.1. Heat flow and heat generation of rocks

Kirovograd block of the USH.

Comparable average TG values for a small number of samples were also obtained for the basement surface of the Carpathians, Crimea and Volyn-Podolsk plate.

Variations of TG at usual sizes of the fields of aged heat generation are not accompanied by a corresponding change in the heat flux. This can be verified by comparing TG and TP within the Kirovograd shield block, where both parameters have been investigated in the most detail (Fig. 3.1).

Changes in heat generation with depth in the consolidated crust are usually estimated from its exponential relationship with seismic wave velocity. Naturally, we mean the correspondence of v_p changes to changes in the composition and degree of metamorphism of rocks with depth. According to the complex of geological and geophysical data, they are described in [3, 10, etc.]. According to these descriptions, it can be assumed that the lower crustal horizons contain up to 30% of ultrabasic rocks in addition to basic granulites. Heat generation in basic granulites is about half of the TG of gabbroids, i.e., about 0.25-0.30 μW/m^3 , in ultrabasites - about 0.04-0.05 μW/m^3 . Thus, above the M section in normal crust (without coro-mantle layer - CM) we can expect heat generation of about 0.2 μW/m^3 . In the upper horizons of the shield crust, the average v_p is about 5.9 km/s. The lowest part of the shield crust (at a depth of about 42-43 km) is characterized by a value of 7.2 km/s. We obtain the expression TG = 1.28 ehr 1.54 (6 $-v_p$). In many cases it is necessary to use density sections of the crust to calculate crustal TG. TG = 1.28 ehr 5.7 (2.69 - σ). Correlations between parameters for rocks of the consolidated crust are obtained for platform temperature distributions. Heat generation in differentially lithified sedimentary rocks is related to v_p as TG = 1.264 - 0.084exp(0.554(v_p - 2)). Differences in depth T suggest a correction to the calculated TGs - accounting for the effect of anomalous temperatures on longitudinal seismic wave velocity (approximately 0.06 km/s per 100^0 C) and density (0.01 g/cm^3 per 100^0 C).

TP and depth T were calculated using expressions for three-dimensional sources (parallelepipeds) with an anomalous temperature ΔT or a given TG level [7, 16, et al.]

The calculations of the crustal radiogenic component of the TP were performed along the profiles on which the velocity and density models were constructed. Their network in Ukraine is quite dense (the total length of profiles is about 10 thousand km) (Fig. 3.2), which allows characterizing the parameter in all major tectonic regions and their large parts. High-quality velocity sections have not been constructed for all HSZ profiles. Accordingly, density models in these cases are less accurate. Therefore, it is necessary to take into account the possibility of a noticeable error in the calculation of the crustal component of the heat flux. It reaches on average 5 mW/m $.^2$

The considered heat generation distributions used to construct the stationary component of the crustal and upper mantle thermal model allow us to explain the observed background TPs at a constant mantle heat flux of about 20 mWm^2 . For the adopted depth interval, these are the minimum T. They are lower only in localized areas of TP distribution below 40 mW/m^2 and, accordingly, crustal rocks with reduced heat generation. If such areas are sufficiently large, the reduced depth temperatures corresponding to them spread to the bottom of the models. There are few such areas in Ukraine and they are shown below.

During model building, the results of temperature calculations were compared with solidus temperatures (Tc) of deep crustal and mantle rocks.

In the granitic and transitional layers of the crust, where rocks of amphibolite facies of metamorphism still occur, melting occurs in the presence of water. Accordingly, at a depth of about 10 km and more, the solidus temperature is about 600^0 C. Toward the surface, Tc rises to 940^0 C. At normal temperature distribution, the bottom of the transition layer corresponds to a velocity of 6.8 km/s and a density of 2.9 g/cm^3 . At heating leading to partial melting, T in the corresponding depth interval of the crust is higher than normal by 200-300^0 C, which lowers the velocity to 6.6 km/s. In the basaltic layer of the crust, melting begins under dry conditions at T around 950-1100^0 C. In the mantle is also considered only

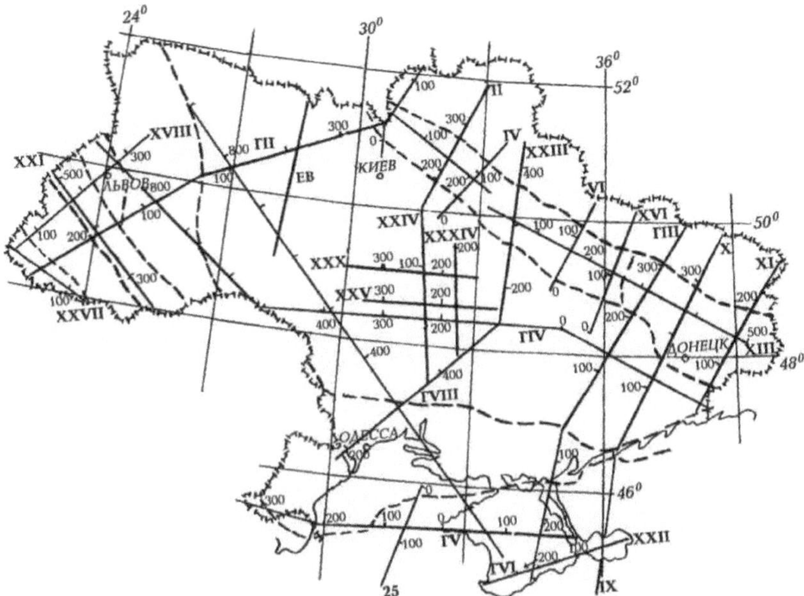

Fig. 3.2. Location of HSZ profiles on the territory of Ukraine.

The index G stands for "geotraverse". EB - profile of the GSS "Eurobridge". Dashed lines are region boundaries (see Fig. 2.1).

dry solidus, since water concentration here is insignificant. The accepted solidus temperature (50 km - 1200^0 C, 100 - 1370^0 C, 150 - 1510^0 C, 200 - 1650^0 C, 250 - 1760^0 C, 300 - 1850^0 C, 350 -193^0 0C, 400 - 1980^0 C, 450 - 2020^0 C) characterizes the very beginning of melting (melt concentration about 1%) "...under the influence of impurity elements..." [27, c. 1265].

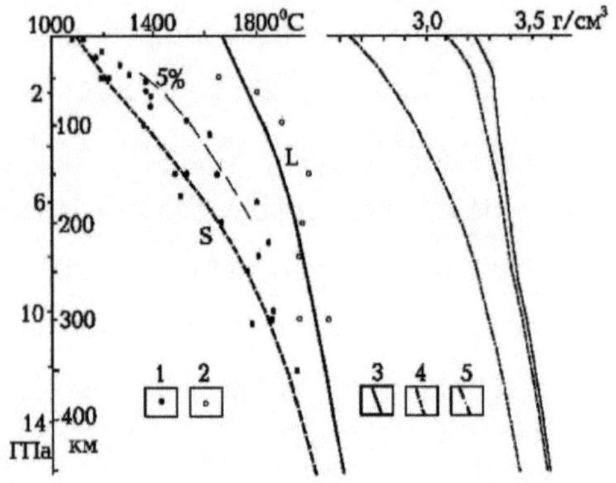

Figure 3.3. Accepted estimates of solidus, liquidus, and density of mantle rocks.

1, 2- experimental data (1 - solidus, 2 - liquidus), 3,4 - densities of mantle rocks (3 - at normal T, 4 - at solidus), 5 - density of basaltic melt.

In Fig. 3.3, it is compared with the liquidus, which has been studied in a greater depth range [42 et al.]. The T_c curve is approximated for the depth range (H) 50-450 km by the expression $T_c = 1013 + 3.914H - 0.0037H^2$. After polymorphic transformation of rocks near the base of the upper mantle, the solidus temperature increases by 200-250^0 C. In the upper interval (at depths of 40-240 km), the solidus temperature is confirmed by the results of an independent study of PT conditions in mantle magmatic sources (up to about 250 km) [13]. Fig. 3.3 also shows the liquidus curve of mantle rocks. It is approximated in the same depth range by the expression $T_l = 1665 + 1.772N - 0.0017N^2$. The estimation of the degree of melting of mantle rocks is significant for the depth interval of 50-200 km. It can be assumed that here by the beginning of the "eutectoid" section melting reaches 5%. Then at increase of T by about 50^0 C there is an increase up to 30-35%.

The information on its density is essential for estimating the movements of mantle matter. They are shown in Fig. 3.3. Apparently, the density of basaltic melt remains less than the density of solid rocks in the whole depth interval under study. But for ultramafic magmas, the situation may be different: already at about 250 km, molten olivine is denser than solid olivine.

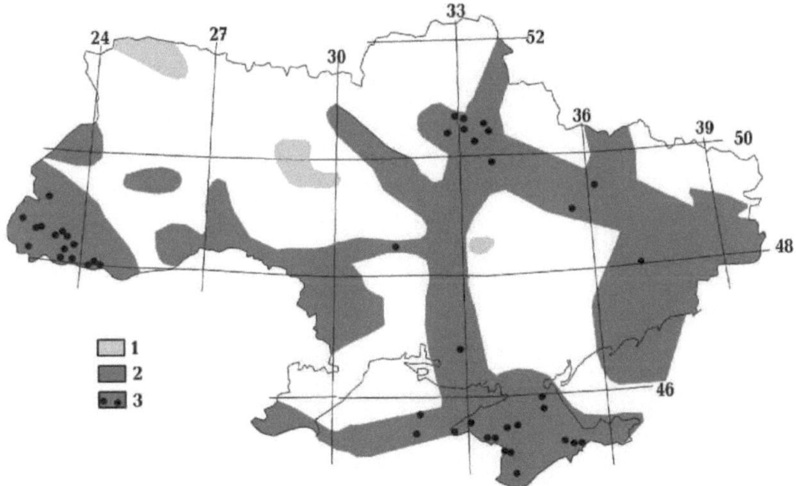

Fig. 3.4. Contours of the areas of reduced TG in the crust (1) and zones of modern activation (2) of Ukraine.

3 - points with anomalous helium isotopy.

The construction of thermal models for unsteady sources is entirely tied to the depth processes occurring within different endogenous regimes according to the advection-polymorphic hypothesis applied by the authors [11 et al]. For models of the Ukrainian tectonosphere, the processes of modern activation (almost in all regions), which started in the mantle several millions, and in the crust - hundreds of thousands of years ago, of the Alpine geosyncline (Carpathians) are important. Some traces in modern temperatures were left by the Hercynian and post-Hercynian processes of the Donbass and the

Cimmerian and post-Cimmerian processes of the Scythian plate. The rather complicated procedure of model building is not considered here. We will only point out the widespread zones of modern activation, the contours of which are constructed according to the data of the complex of geological and geophysical methods (Fig. 3.4).

As can be seen from Fig. 3.4, modern activation also occurs within the alpine Carpathian geosyncline.

3.2. Three-dimensional state-of-the-art thermal model

Temperatures associated with different-age unsteady sources in the crust and upper mantle were added to those caused by the

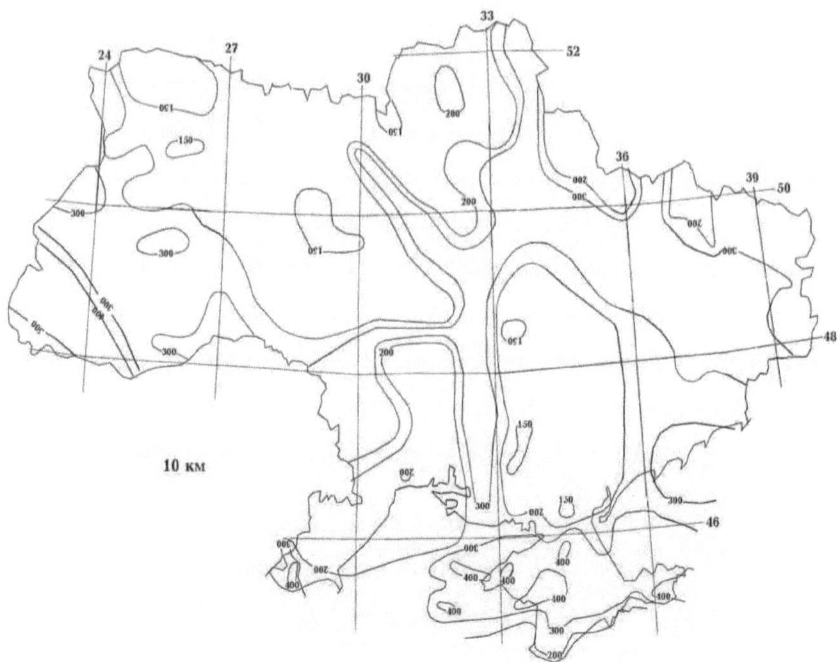

Figure 3.5. Temperature distribution at a depth of 10 km.

practically stationary sources and form a modern three-dimensional thermal model. It is represented by several slice maps for the territory of Ukraine (Figs.

3.5-3.8).

Temperature differences at the same depth between the platform regions with practically stationary field and active zones sharply increase with approach to the centers of nonstationary heat sources in the interior of the zones of modern activation and alpine geosyncline. They reach at 10 km depth 350, 25 km - 450, 50 km - 650 and 75 km - 900^0 C. This list does not include small areas with reduced crustal heat generation (see Figures 3.5-3.7). The degree of their difference from areas with normal background platform temperatures becomes smaller at greater depths. The maximum T values are reached in the interior of the Alpine geosyncline of the Carpathians, especially in its inner part - the Transcarpathian Trough.

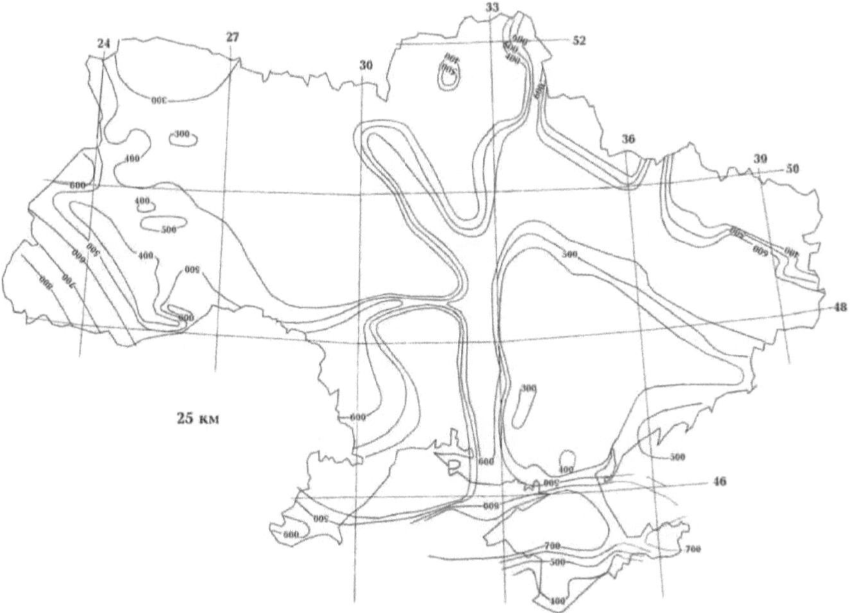

Figure 3.6. Temperature distribution at a depth of 25 km.

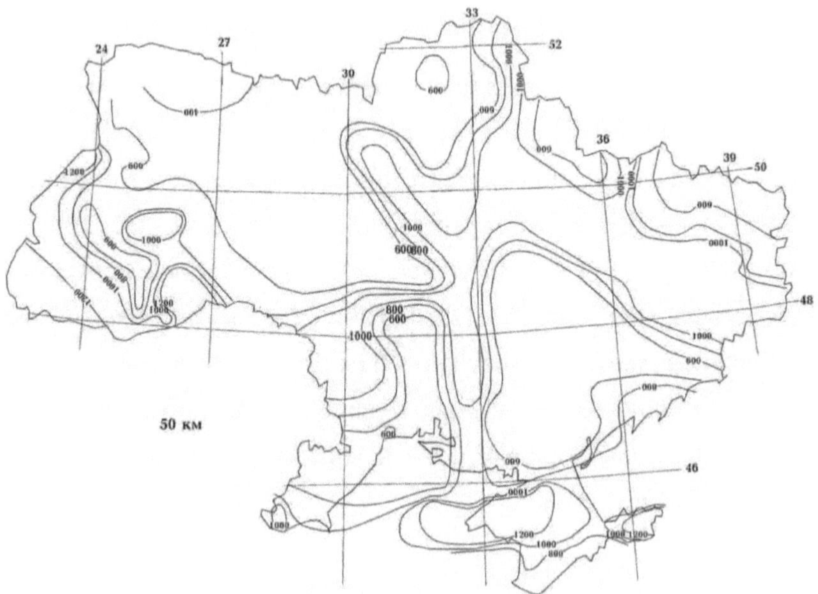

Figure 3.7. Temperature distribution at a depth of 50 km.

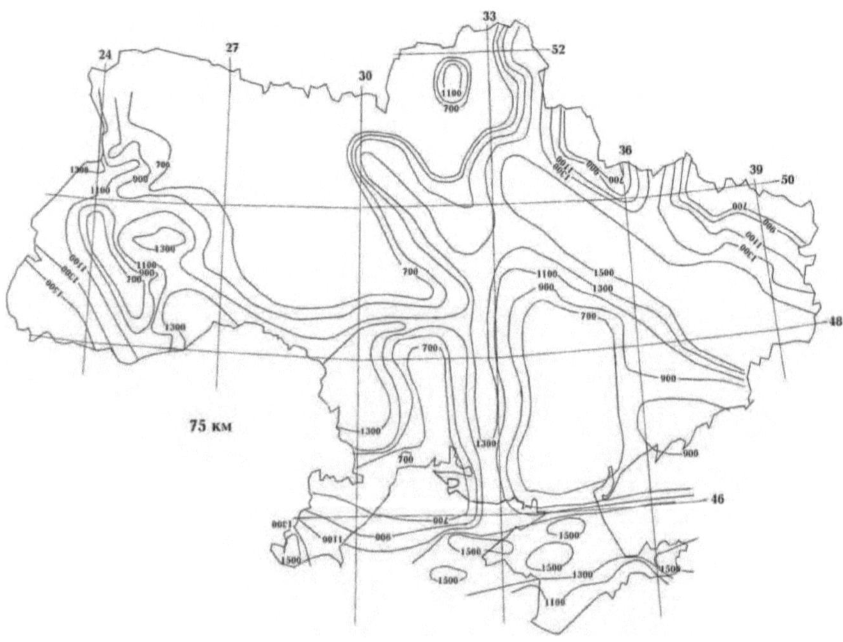

Figure 3.8. Temperature distribution at a depth of 75 km.

Comparison of depth T with solidus temperatures shows that zones of partial melting are common not only at maximum depths in active regions, but also in some places in the crust.

3.3. Model validation

The most obvious way to test this is with geothermometers, but it only works well for the crust and platform mantle.

The calculated temperatures are confirmed by independent experimental data. But anomalous T of the mantle are not reflected in the latter.

Under the conditions of temperatures and pressures of the upper mantle, the preservation of mineral associations that arose under active regimes (in the particularly at elevated temperatures) is limited in time. B

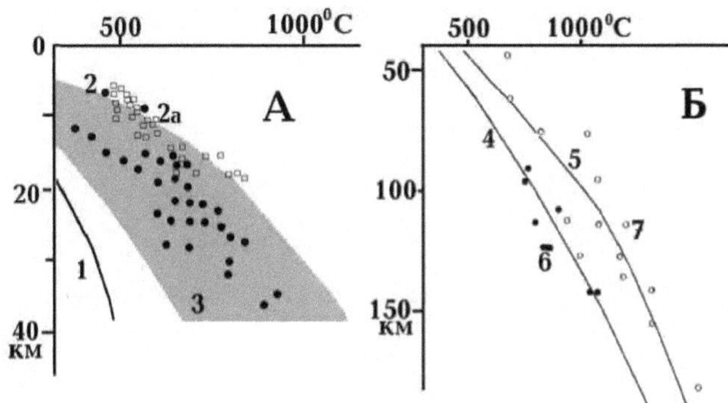

Fig. 3.9. Thermal models of the crust (A) and upper mantle (B) of Ukraine and geothermometer data.

1-calculated thermal model of the platform crust at normal heat generation of rocks, 2 - data on RT-conditions of crustal rocks formation, opened by erosion on the USH, 2a - on the Rakhiv and Marmarosh massifs of the Carpathians, 3 - calculated range of crustal T at activations (including T at the margins of active zones), 4,5 - calculated thermal models of the platform mantle (4 - at reduced TG, 5 - at normal TG), 6,7 - RT-conditions of formation

rocks brought to the surface by kimberlite magmas (6 - in the area with reduced

TG - on the Pripyat Shaft, 7 - in the area with normal TG - on the Ukrainian Shield).

Unlike the crust, where PT conditions allow associations to persist until the emergence of a regime with higher T, mantle rocks return to the platform mineralogy after a sufficiently long time without activation, i.e., the exported xenoliths can serve as geothermometers only for the platform regime. Outside Ukraine there are areas with other data on xenoliths-geothermometers in the mantle. In our case, it is necessary to refer to the data on PT-conditions in magmatic centers arising during the periods of geosynclinal development of the Carpathians, Donbass, Crimea and rifting - DDV. Average temperatures at a depth of about 50 km 1250^0 C and at 75 km - 1350^0 C have been established. Deviations from the given T - on average about 50^0 C, there are also deviations by 100^0 C. I.e., we can state that the calculated temperatures in the mantle and in the active regions are real.

A peculiar indicator of the region's activation (and, accordingly, of the anomalous thermal model of its subsoil) is oil and gas content. It manifests itself, of course, only in zones where there is material (carbon, hydrogen "provides" the activation process itself) for the formation of a significant amount of hydrocarbons [12]. All oil and gas bearing areas of Ukraine fall into the zones of modern activation.

The existence of the mantle floor of the active deep process is indicated by anomalous helium isotopy [21]. It has been recorded in the Carpathians, MDV, in the USH, in Donbas, on the South Ukrainian monocline, and on the Scythian plate. The observation network is sparse and irregular, but still we can speak about the detection of the mantle floor of partial melting in most of the activation zones (Fig. 3.4).

Geophysical control of the model is associated with the identification of objects with anomalous physical properties of rocks, which arose under the influence of high temperatures, partial melting, fluidization of part of the section above the melt layers.

In the most complete form, it can be carried out on the basis of the data of the universally studied gravitational field. The influence of high temperatures of the subsurface is clearly manifested in this case when considering the mantle gravity anomaly [9, 11, etc.] - the difference between the observed field and the effect of the crust and normal mantle. In all zones of modern activation on the platform, the predicted (calculated from decompaction and compaction from T anomalies) perturbations of minus 20-30 mGL intensity are observed, in the Donbass they are somewhat higher (up to -40 mGL), on the Scythian plate they are even higher due to the influence of decompacted sources in the Black Sea mantle, and in the Carpathian region they reach -200 mGL (the influence of anomalies that arose during the Alpine geosynclinal process).

Anomalous conductive objects in the middle and upper parts of the crust, associated with the rise of fluids above the melting zones, are noted in all geoelectrically studied zones of modern activation (almost all, presented in Fig. 3.4). The calculation shows their correspondence to the concentration of formed fluids. When coinciding with graphitization zones, the conductivity increases sharply. It is less often possible to study mantle conductors associated with the same process and corresponding to partial melting of the subcrustal mantle. But nevertheless, such data are not uncommon in the USH, Donbass, Carpathian region, and Scythian plate.

Seismic wave velocity anomalies corresponding to overheating of parts of the crust and upper mantle in the activation zones are identified sporadically in the Carpathians, Scythian plate, DDV and Donbass. Their intensity corresponds to the model temperature anomalies.

The sum of the above information allows us to state that the control of the three-dimensional thermal model of the crust and upper mantle horizons of Ukraine was successful.

Chapter 4. Geothermal Resources of Ukraine

The importance of terrestrial heat in the energy balance of the world is still insignificant. But it is the fastest growing part of the energy sector [26, 34, etc.]. Introduction in recent years of new technologies of heat extraction (heat pumps, etc.) [34, 48, 51, etc.] demonstrates the possibility of geoenergy to take one of the leading places in the utility sector of the industry. In developed countries, about one million geoenergy plants for homes have been commissioned in the last few years. The environmental side of things is also attractive: modern geoenergy systems provide for the complete return of deep water to the reservoir. Therefore, the analysis of geothermal data from the point of view of assessing the resource potential of thermal energy seems to be important and relevant.

This paper deals with studies of a regional nature aimed specifically at estimating resource density (W). Although the transition to the determination of reserves of deposits is already feasible at the present time with the emergence of specific tasks in a number of regions of Ukraine. In accordance with the requirements developed for other minerals, such estimation can be performed in different variants [24, 25, 38, 43, 44, etc.]: with different degrees of validity and with orientation on different technologies of heat extraction. The most acceptable (fully reflecting the energy potential of the region) seems to be the circulating technology of heat extraction from dry rocks [38, 40, etc.]. Calculations will be carried out for it, which can be revised, if necessary, taking into account the requirements of other technologies.

According to the degree of validity, resources are usually divided into prospective (C3) and inferred (P1 and P2). When assessing resources of P2 category, only the possibility of the presence in the region of conditions for the formation of geothermal energy deposits is considered. Information on the temperature distribution in the subsurface is obtained on the basis of geological and geophysical data (only partially - geothermal data), the maximum achievable drilling depth (10 km) is introduced in the calculations. It is assumed that the rock massif can be cooled down to the surface temperature. Obviously, only a maximum estimate is feasible in this way, which is of little use for identifying specific areas potentially promising for Earth's heat extraction. In the assessment of P1 category resources, regions are studied for which the possibility of energy extraction is already clear in principle. Calculations are carried out for real drilling depths (up to 6 km) and the requirements of different energy consumers for the temperature of the coolant inlet to the heat exchanger and its discharge are taken into account. Prospective C3 resources also take into account the economic feasibility of utilizing the earth's heat, which is expressed in limiting their densities, at which the obtained energy can compete with the energy supplied by traditional sources.

The boundary between P1 and C3 resources shifts with changes in technology and the cost of energy from traditional sources. Therefore, the authors aimed to carry out calculations for the entire territory of Ukraine, stipulating the level of W, reflecting the current position of the boundary between P1 and C3. In this case, we will focus on resources suitable for use in heat supply, i.e. for extraction from the geocirculation system (GCS) of water at a temperature of 60^0 C and its discharge at 200C. These are the maximum resources, as for heating and electricity generation

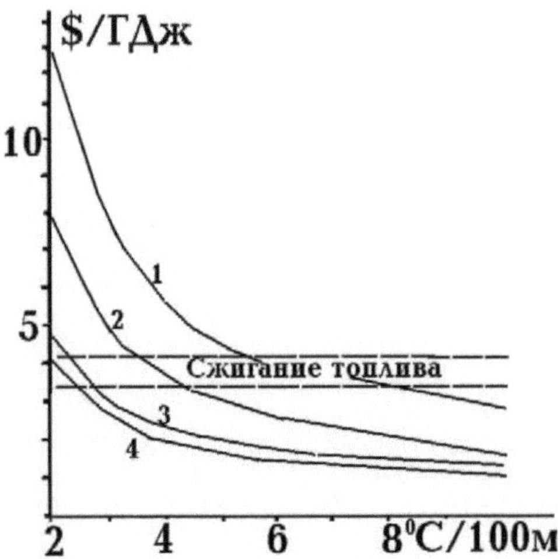

Fig. 4.1 Cost of sales

products of heat supply GCCs depending on geothermal conditions and technology level. Economic model of MTI 1990 [51].

1-4 are variants of the HCC technology.

(steam for turbines) need 100-40^0 C and 210-70^0 C, respectively. The approach adopted allows the use of internationally recognized results of economic evaluations performed at the Massachusetts Institute of Technology, shown in Fig. 4.1. They indicate the achievement of profitability of geothermal energy production by GTS for the most advanced technologies at the level of geothermal gradient (γ) of 2.0-2.5^0 C/100m. An example of practical use of thermal energy in the area with such a value of γ is also available in Ukraine.

One feature of a real geothermal energy source that is often misrepresented should be noted. It is about its categorization as renewable. In principle it is true: the heat extracted from the subsoil will be compensated by the heat coming from below. But the rate of such renewal with real thermal properties of the environment is incomparably greater than the known history of mankind, i.e. practically zero.

4.1. Calculation methodology

The calculation of the thermal resource density is performed as follows [25 et al:]

$$W = N \cdot K \cdot C\rho \, \Delta T(H_з - H_в)$$

where N is the norm of fuel consumption for commercial heat - $0.34.10^{-10}$ t c.u./J (t c.u.t. - ton of conditional fuel: in 1t of oil - 1.47 t c.t., in 1t of hard coal - 0.9 t c.t., 1t of condensate - 1.54 t c.t., $1000m^3$ gas - 1.25 t c.t., 1 t of lignite - 0.49 t c.e.), K - temperature extraction coefficient (accepted in [25] as 0.125), *Cp* - volumetric heat capacity of rocks, it can be considered almost constant - 2.5.106 J/m^{3} ^{0} C, Δ T - difference in the temperature of the coolant and discharge - 40^0 C, N_z - depth of the bottomhole at which the lower T is determined. Accordingly, W = 0.000425(N_z - N_v) in t u.t./m^2 at H in m.

Depth H_v is the temperature at which provides the average T in the interval $H_з$ - $H_в$, equal to 600C. It is defined as ($T_з$ - $T_т$)/0.5 *γ*, where T_t is the coolant temperature, *γ is the* average geothermal gradient in the interval.

At high T at the lower point it turns out that the upper point falls above the surface. In order to avoid this situation, a restriction is imposed on the T at the upper point: it must be 10^0 C above the discharge water temperature, i.e. 30^0 C. In this case, the difference between the average T of the produced water and the standard value of 60^0 C must be taken into account. This creates an additional multiplier in the formula for calculating W, which is (T_{cp} -20)/40.

Thus, the task is reduced to calculation of T for a given region (given distribution of thermal conductivity with depth) at different real for the region depth heat fluxes and subsequent calculation of W for drilling depth of 6000 m (calculations for 4500 and 3000 m were also carried out). Taking into account the specific surface temperature at the place of deep T calculation gives variations of W values up to ±4 % (for example, when 8^0 C is replaced by 6 ÷ 10^0 C). Therefore, in principle, it is possible to introduce one T0 in each region when calculating T by GTP.

Obviously, the temperature extraction coefficient is not a constant. It must be determined based on the actual conditions of the procedure under consideration.

The calculation shows that it makes sense to take into account not all process parameters. First of all, the temperature anomaly for the whole time of the system operation does not noticeably go beyond the fractured zone. The time of system existence may be a limitation. It is connected with siltation of the fractured zone around the well, because of which the re-injected water is no longer assimilated in the required amount, energy consumption for operation of injection pumps increases, etc.. According to the known data, it is possible to accept the duration of operation of HCC of 25 years. If it is not exceeded, the moment of heat extraction cessation is the moment when the average temperature in the reservoir from which water is pumped out reaches 60^0 C.

Let us determine the operation time. According to [51 et al.], the thickness of the layer in which fracturing is created can reach 500 m. We estimate the size of the fracture area in plan as 250x250 m. The bound porosity is approximately 0.1 of the rock volume. Its change does not noticeably affect the result: depending on the porosity, the filling of the system at the same injection pump capacity will be more frequent (with a smaller unit heat effect) or rare (with a larger unit effect).

The amount of injected water at acceptable energy costs for pumping is 1000-7000 m^3 /day. At real value (simplifying the calculation) 4300 m^3 /day the porous system is filled in 2 years. The decrease of T in volume is T_{a1} = 0.167dT (taking into account the ratio of volumetric heat capacities of water - $4.18.10^6$ $J/^0$ $S.m^3$ - and rock). dT is the difference between the average T at depths of 5500-6000m and 20^0 C. The calculation shows that the resulting anomaly is completely preserved in the object (not only for two years, but for the entire real period of calculations). Therefore, T_{a2} = 0.167(dT - T_{a1}) and so on until (dT -

sum of T_{ai}) is 400C.

The estimated time for a drilling depth of 6 km is 8-23 years for geothermal gradients of 2-6^0 C/100 m, and 1.5-13 years for 3 km. Thus, the system lifetime is not exceeded and K can be defined (T0=10^0 C) as (5750γ - 50)/(6000 - 20/γ)6γ, where γ- in^0 C/m. K is at γ= 0.02 - 0.108, 0.03 - 0.127, 0.04 - 0.136, 0.05 - 0.141.

In the regions, the depth temperatures were calculated from the depth TP values (the observed values are not suitable for this purpose when we are talking about large depths) for the stationary distribution, and the corrections were excluded from the calculated T values. In the Dnieper-Donets Basin (DDV), it was more convenient not to exclude the hydrogeological correction, but to introduce a slightly increased thermal conductivity in the upper part of the section.

The values of average effective thermal conductivities (λ) of rocks in the depth intervals 0-1.5, 1.5-3, 3-4.5, and 4.5-6 km given in Table 4.1 (in W/m·^{0} C) were used for calculations.

Table 4.1.

Δ H, km	11	10	9-10	9	7	1и2
0-1,5	1,85	2,65	2,45	1.8	2.1	2,65
1,5-3	2,65	2,65	2,45	2.25	2.65	2,65
3-4,5	2,65	2,65	2,45	2,65	2.65	2,65
4,5-6	2,65	2,65	2,45	2,65	2,65	2,65
0-6	2,39	2,65	2,45	2,28	2,49	2,65
Δ H, km	1,2-slopes	3-board	3	4	5	6
0-1,5	1,7	1,8	1,8	2	1.8	1.6
1,5-3	2,65	2,05	2,05	2.1	2.2	2.05

3-4,5	2,65	2,65	2,2	2.3	2.65	2.5
4,5-6	2,65	2,65	2,3	2,5	2,65	2,65
0-6	2,32	2,22	2,07	2,21	2,27	2,12

Region numbers: 11 - Transcarpathia, 10 - Folded Carpathians, 9 - Pre-Carpathian Trough, 7 - Volyno-Podolsk Plate, 1 - Ukrainian Shield, 6 - DDV, 2 - slope of the Voronezh Massif, 4 - Donbass, 5 - South Ukrainian Monoclinal, 6 - Scythian Plate.

The proposed method of calculating the depth T contains obvious sources of errors, first of all - failure to take into account the real values of thermal conductivity at the calculation point. Therefore, for all regions the calculated and measured T at the maximum measurement depths were compared. Exceptions were the USH and its slopes, where there are practically no deep boreholes (except Krivorozhskaya and Kirovogradskaya with depths of 5 and 3 km, respectively). The constructed histograms reveal modal values of deviations in the Carpathians, Transcarpathia, Predocarpathia and Donbass about 1^0 C, in the DDV, Crimea, Volyno-Podolsk plate and South Ukrainian monocline - 3-4^0 C. Differences sharply increase only in areas with thick salt beds, but they are unsuitable for the creation of HCS. Thus, the errors of T calculation cannot noticeably affect the determination of W value: the predicted error is up to 10%, which is comparable to the error of GTP.

Let us determine the W level limiting the areas with resource density category c3. For the minimum geothermal gradient of 2^0 C/100 m it is 2.5 t c.t./m^2 . Table 4.2 summarizes the GTP values in different regions of Ukraine that correspond to this and other W values. Obviously, the relationship between the density of geothermal resources and the value of the deep heat flux is quite complicated, especially for large values of W.

Table 4.2.

W, t c.t./m^2	TP, mW/m^2 in regions											
	11	10	9 10	-9	7	1и2	1,2-slopes	3-boards	3	4	5	6
3		59	54	50	55	59	51	49	46	49	50	
4		69	64	60	65	69	60		54	58	59	56
5	73	80	74	69	75	79			62	67	68	64
6	82			79						76		73
7	92									85		81
8	10											
9	11											
10	12											
TPSzmin	48	53	49	46	50	53	46	44	41	44	45	42

Obviously, resource densities meeting the c3 category are fairly widespread.

It is of some interest to compare W values with data on hydrocarbon fields. Let us consider the density of energy reserves, which can be obtained in the form of marketable heat from a large oil field in the Far East (without taking into account energy costs for oil transportation and with efficiency of conversion into useful heat of 0.8). Let's assume such real parameters of the field: thickness of productive layer - 180m, porosity of reservoir rocks - 0.15, pore filling factor - 0.75, recoverability factor - 0.37, oil density - 0.8 t/m^3 . We get 8.8 t c.t./m^2 . At a shallow field (which in the conditions of Ukraine is considered profitable to exploit in the presence of ready wells) the density of reserves is an order of magnitude less.

Thus, even in terms of concentration, geothermal energy is comparable in some areas to that concentrated in traditionally used hydrocarbon fields. Its distribution area is incomparably larger.

The above calculation of K assumes a "one-time" heat extraction methodology. In this sense, the value of W (w_6) seems to be sharply underestimated. It is obvious that energy extraction can be continued even after exhaustion of its source at a depth of 5.5-6 km (possibly, without drilling additional wells). It seems possible to obtain energy from depths of at least 2.5-3 km at a geothermal gradient of 2^0 C/100m. Carrying out corresponding calculations for other depths of the bottom of the exploited interval (H, in km), we obtain values W = (0.427H - 0.07)(γ - 2.7 + 0.3H). For example, at a typical heat flux of 45 mW/m^2 , the "full" value of W will increase compared to W6 by a factor of 4.5. Note, by the way, that using data on regions characterized by different values of w_6, it is easy to obtain (for the range W6 2.5-10) W3 = 0.53(W6 - 1.5) and $W4_{.5}$ = 0.78(W6 -0.8).

4.2. Initial data

In the considered case, the calculation of geothermal resource density was based on the values of pre-determined heat flux. The study of the territory of Ukraine (achieved mainly by the efforts of the authors) on this parameter is unique. Therefore, it is here it is convenient to demonstrate the potential possibilities of utilization of the Earth's heat in the regions, most of which are not distinguished by a particularly large energy potential.

In addition to the large amount of heat flux information, Ukraine is distinguished by the use of depth values of TP. This term implies the introduction of corrections into the observed values to account for the influence of near-surface distortions (see Ch. 1).

Due to the regional nature of the study, GTP values in wells that were close to each other (within one minute of latitude and longitude) were averaged. As a result, 5600 points were plotted in the schemes below. The density of the network of GTP determinations is very uneven. It is clear that the majority of values were obtained in oil and gas bearing and coal-bearing areas and in localized areas of ore fields. In other places (first of all, in the most part of the USH and its slopes, slope of the Voronezh massif, part of the Folded Carpathians) "white spots" are widespread.

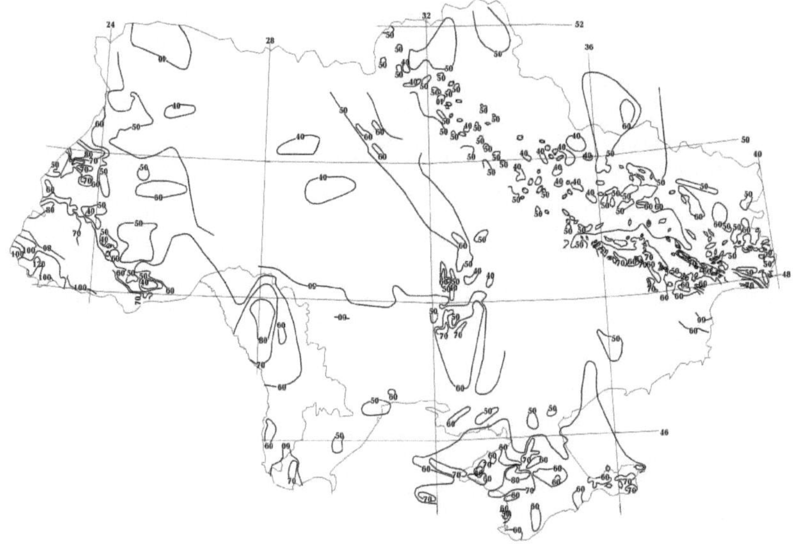

Fig. 4.2. Depth heat flux in the territory of Ukraine and Moldova.

The distribution of GTP is presented in Fig. 4.2. This version of the map is the last, it is somewhat more complete than the one presented in [36].

The difference between the maximum GTP values in the Transcarpathian Trough (120-130 mW/m^2) and the minimum values in the Ukrainian Shield (30-35 mW/m^2) reaches 4 times, the calculated W values differ even more (see below). Even before calculating the density of geothermal resources, it can be assumed that they are predominantly concentrated in three large basins: western, southern and eastern, separated by an area in the center of Ukraine with minimal resources.

The detailed study of most regions of Ukraine allows us to reveal another feature of the field, which is not diagnosed with a sparse network of measurements. It is about local anomalies. Most of them cannot be shown on the maps below, but it is within them that the maximum geothermal energy in the region can be obtained (Fig. 4.3).

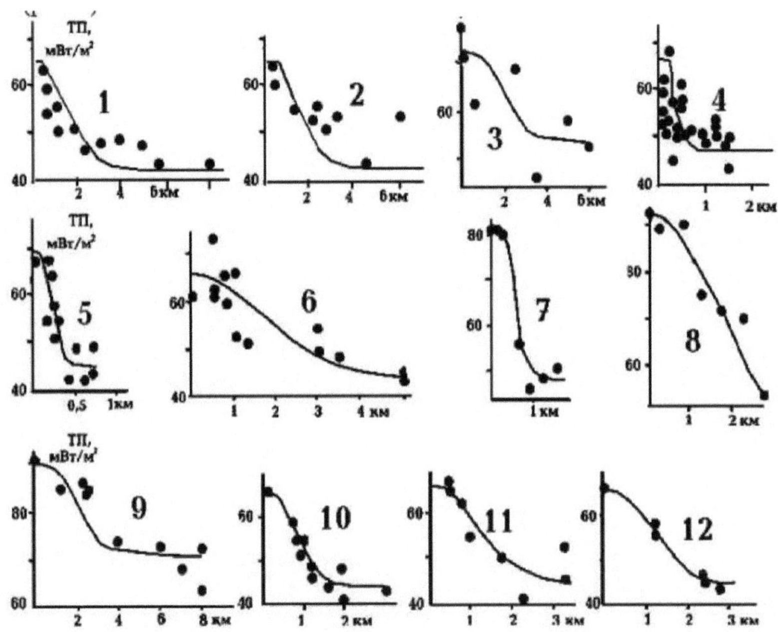

Fig. 4.3. Local anomalies of deep TP in different regions of Ukraine.

Regions (studied areas): Volyno-Podolsk plate (1 - Lokachinskaya, 2 - Velikomostovskaya, 3 - Yavorovskaya), Precarpathian trough (4 - Letnyanskaya), Ukrainian shield (5 - Yurievskaya, 6 - Kirovogradskaya), Donbass (7 - Mikhailovskaya, 8 - Konstantinovskaya), Scythian plate (9 - Novoselovskaya), Dnieper-Donets depression (10 - composite anomaly for several areas of the central part of the DDS, 11 - Malodevitskaya, 12 - Yablunovskaya). Dots - experimental GTP values, lines - calculated.

The transverse size of the anomalies is the first kilometers, the intensity of the disturbance (exceedance over the local background) is rather uniform - about 20 mW/m^2 , which corresponds to an increase in W (see Table 4.2) by about 2 t.u.t.m^2 . Geological data and special calculations show that the anomalies are confined to areas where heated fluids approach or come to the surface, the source of which is located at a depth of 6-7 km in zones of modern activation

[1, 16, 22, 43, 44, etc.].

The width of the permeable zone, through which fluids move, appears to be small - at the level of the first hundreds of meters. Regardless of geothermal data, the same model was built by A.E. Lukin on one of the oil and gas bearing structures of the Far East [33] - Fig. 4.4. It is noteworthy that the parameters of the circulation system are close in all cases, despite the tectonic heterogeneity of the studied regions.

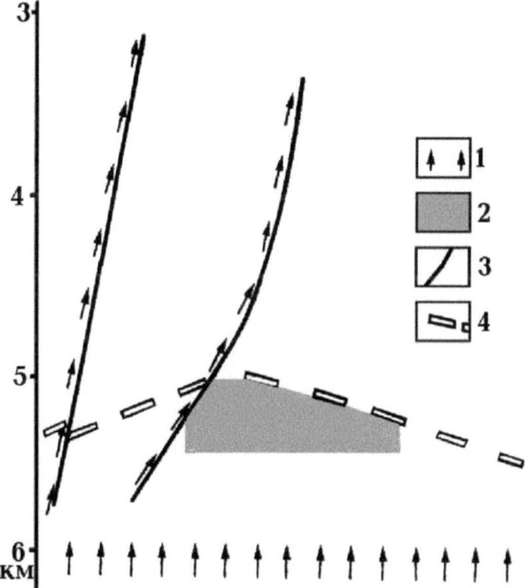

Fig. 4.4. Injection of deep water along disturbances in the Machekhskoye field (according to [33] with simplifications).

1- direction

movements of high-pressure deep fluids, 2 - gas deposit, 3 - disturbance, 4 - unconformity surface (screen?).

Thus, already at the reached stage of regional studies we can talk about the detection of individual

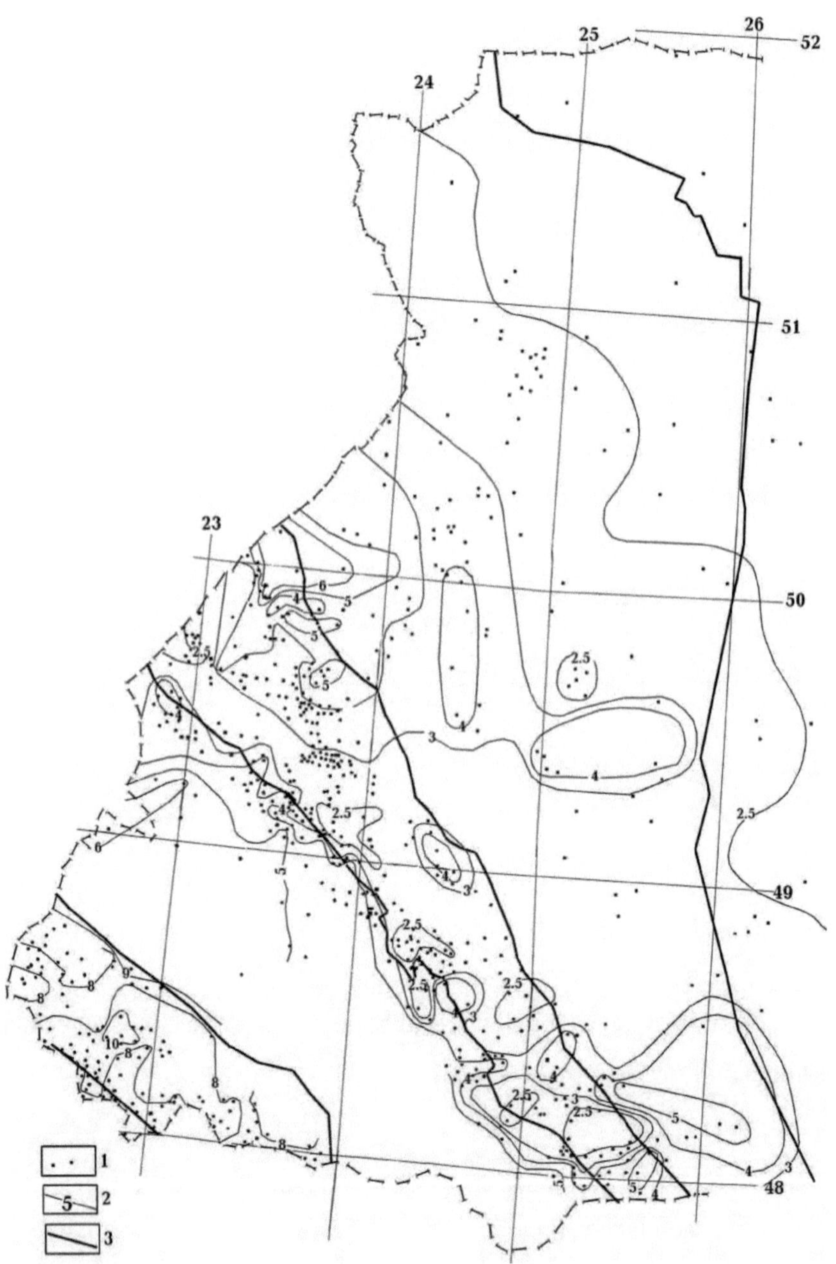

Fig. 4.5. Geothermal resources of western Ukraine.

1 - points of W values determination, 2 - isolines of W_6 (in t.u.t./m^2), 3 -

boundaries of tectonic units. From southwest to northeast: margin of the Pannonian Depression, Transcarpathian Trough, Folded Carpathians, Precarpathian Trough, Volyno-Podolian Plate.

geothermal energy deposits, which are found practically in all regions of Ukraine, including the Ukrainian Shield, the prospectivity of which for this type of mineral is generally low.

The above data emphasize the need for special geothermal studies aimed precisely at studying the energy potential, and not only at solving the problems of regional geological and geophysical study of the Earth.

4.3. Distribution of geothermal resources

The distribution of geothermal resources in the western basin is presented in Fig. 4.5.

On the Volyn-Podolsk plate, the W6 level is determined by the transition from low values of the shield slope to elevated values in the Carpathian geosyncline, which is at the stage of postgeosynclinal activation. In the northern part there is a zone of extremely low W (much less than cost-effective) in place of the negative Volyn heat flow anomaly. The average level of geothermal energy concentration is low - from 1.5-2 to 2.5 t eq/m^2 . Elevated W6 values are observed only within the Yavorivka, Ternopil and Chernivtsi heat flow anomalies (up to 4-5 tce/m^2). These disturbances are confined to zones of modern activation, where other geological and geophysical signs of such a process are also noted.

Approximately the same picture was obtained for the Precarpathian trough (laid down - except for the southwestern margin - on the Precambrian basement), where the typical values of heat flux are small and the growth of TP is noticeable only within the western parts of the Yavorivka and Chernovtsy anomalies and on the border with the Folded Carpathians.

In the Folded Carpathians, the thermal field has been studied mainly in the

Skib Zone, partially overlain by the advanced trough. In the main part of the region, the studied wells are rare, forming a large "white spot (Fig. 4.5). W6 values average 3.5 t c.e./m^2 , despite a rather high GTP. This is due to the significant value of rock thermal conductivity, which lowers the geothermal gradient.

In the Transcarpathian trough, W6 values are maximum for the territory of Ukraine. In some areas here it reaches 10 tce/m^2 . Naturally, this region seems to be the most promising for utilization of the Earth's heat. Hot water is extracted here for balneological purposes, for heating, construction of a geothermal power plant was planned. However, the structure intended for it is currently being used as an underground gas storage facility for a transit gas pipeline.

In the total amount of resources of the western basin, the Volyno-Podolsk plate plays quite a significant role (despite the universally low W6) due to its large area. A total of 0.25 trillion tons of fuel equivalent has been accumulated in the basin (we are talking about the depth interval of 5.5-6 km, the resources of which can be significantly supplemented by those located at shallower depths - see above).

In the southern basin (Fig. 4.6), during the transition from the slope of the Ukrainian Shield to the South Ukrainian monocline and then to the Scythian plate, there is a gradual increase in the W6 value from north to south from 2.5 to 3.5-4 tcf/m^2 . In the Crimea, North Dobrudja and Pridobrudja trough, anomalies with intensity up to 7 tce/m^2 in the zones studied by the complex of geological and geophysical methods are distinguished.

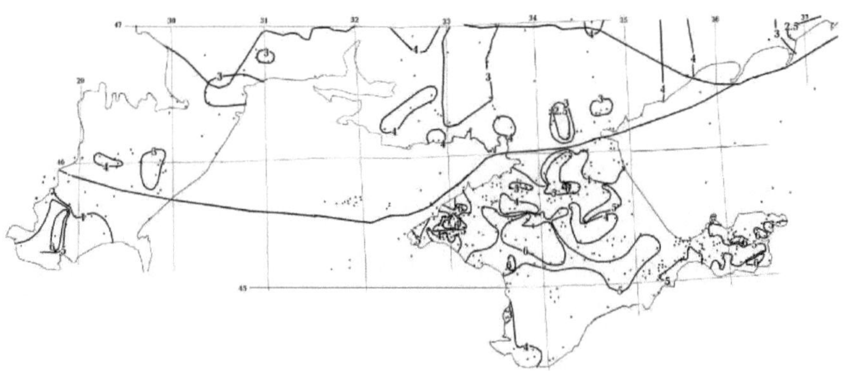

Fig. 4.6. Geothermal resources of southern Ukraine.

See Fig. 4.5 for reference designations. 4.5.

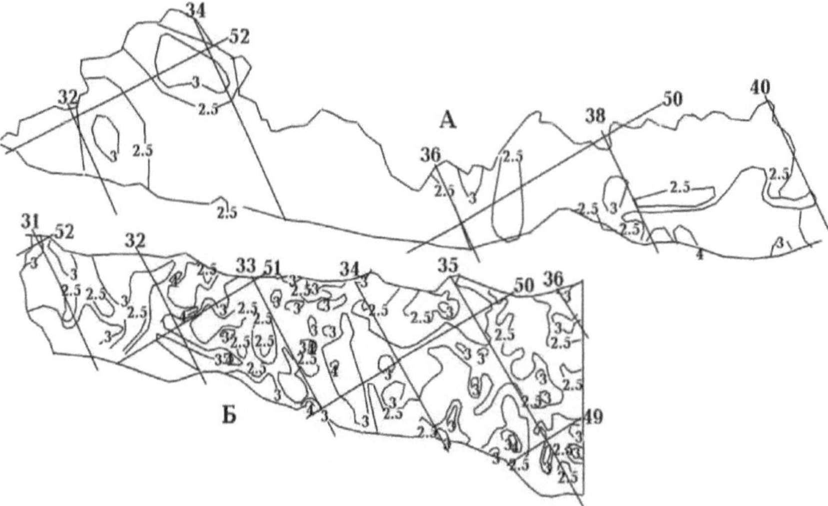

Fig. 4.7. Geothermal resources of the Voronezh massif slope (A) and the Dnieper-Donets depression (B).

See Fig. 4.5. 4.5. of modern activation. For some of them, it was established that the warming of several km thick strata by hot deep fluids was confined [16 and others].

The total amount of geothermal resources in the basin is quite significant due to its large area - 0.3 trillion tce.

In parts of the eastern basin, the study of the heat field is fundamentally different. In Donbass, it is maximum (6500 single heat flux determinations), and on the slope of the Voronezh massif it does not exceed that achieved on the Ukrainian Shield. Therefore, the parts of the basin are presented on a different scale (Figs. 4.7 and 4.8).

On the slope of the Voronezh massif, data for the territory of Russia were also used in constructing the scheme of W6 distribution. This only partially changed the possibility of identifying areas with profitable W values, "white spots" are distributed. Only on the border with Donbass a small area with values of the parameter more than 4 t c.u.t./m^2 appears.

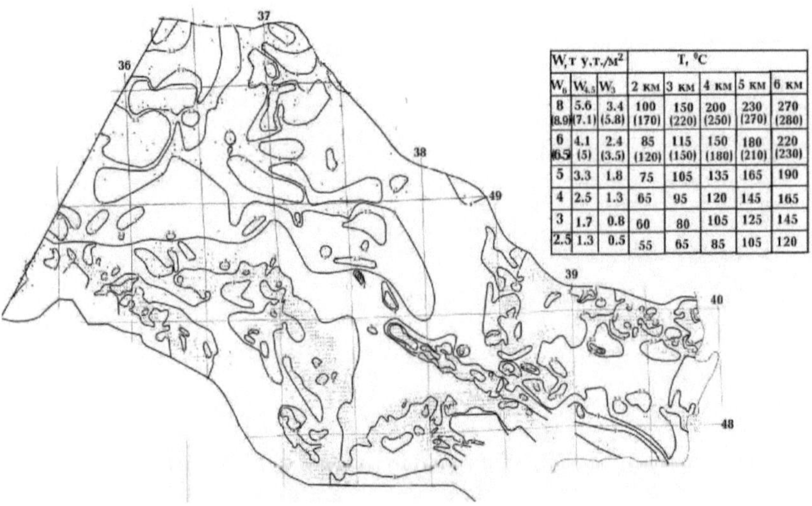

W, т у.т./м^2			T, ^{0}C				
W_6	$W_{4.5}$	W_3	2 км	3 км	4 км	5 км	6 км
8 (8.9)	5.6 (7.1)	3.4 (5.8)	100 (170)	150 (220)	200 (250)	230 (270)	270 (280)
6 (6.5)	4.1 (5)	2.4 (3.5)	85 (120)	115 (150)	150 (180)	180 (210)	220 (230)
5	3.3	1.8	75	105	135	165	190
4	2.5	1.3	65	95	120	145	165
3	1.7	0.8	60	80	105	125	145
2.5	1.3	0.5	55	65	85	105	120

Fig. 4.8. Geothermal resources of Donbass.

See Fig. 4.5 for reference designations. 4.5.

In the Dnieper-Donets Basin, it is obvious that there is a fairly significant distribution of areas with promising geothermal resources. However, the usual concentrations of geothermal energy in the Dnieper-Donets Basin are small - about 2.5-3 tce/m^2 . Only rarely are there areas with W6 greater than 4 t eq/m^2 (exceptions are discussed above). In this case, the boundary between the DV and Donbass is drawn rather conventionally: a rather extensive transition area is replaced by a straight line, approximately separating the territory with a denser observation network in the Donbass mine fields from a noticeably less dense one in the hydrocarbon fields in the Dneprovsk-Donets Basin (Fig. 4.7).

In the Donbass, the concentration of geothermal energy is much higher than in the DDV (Fig. 4.8). However, here the main increases in the values of calculated W6 are associated with the movement of water along permeable fault zones. The temperature distribution in them with depth (up to 6 km) is practically stationary, although it should differ from that calculated using the formula used. It is not difficult to refine it, but this operation makes sense in more detailed studies. Radical changes in W6 are unlikely to occur in this case. The subparallel strike of faults near the axes of the Main and Druzhkov-Konstantinovskaya anticlines probably leads to a significant extension of thermal anomalies. Therefore, they are manifested on the constructed map (Fig. 4.8).

The average concentration of geothermal energy in Donbass is about 4-4.5 t c.u.t./m^2 , increasing in the northwestern part of the Main anticline and in southwestern Donbass to 6-7 t c.u.t./m^2 . W6 anomalies are quite widespread in the region. It should only be noted that the values of the calculated parameter in its southwestern part with low thickness of the sedimentary cover are somewhat overestimated.

Fig. 4.9. Geoenergy resources of the Ukrainian Shield and its slopes.

See Fig. 4.5 for reference designations. 4.5.

The calculations here used the same λ value as in other areas of the region. In fact, a significant depth interval is represented by crystalline basement rocks with average thermal conductivity 15-20% higher. The comparison of observed and calculated temperatures does not reveal the allowed error, because the depth of the boreholes in which T was measured is not large enough (about 1 km).

The sum of geoenergy resources in the eastern basin is about 0.4 trillion tons of fuel equivalent.

As noted above, the poor study of most of the SC does not make it possible to characterize the thermal field over significant areas of the SC. This, of course, also applies to the distribution of W. The data presented in Fig. 4.9 show that areas with promising c3 resources may occur on the shield and its slopes, but their identification and study are still to be carried out in the future. Extending the idea of low heat flux to the entire shield beyond the Kirovograd anomaly (and some other minor TP disturbances), we can estimate the average W6 value of 1.8 tce/m ^{2}

To the northeast of the shield (already within Belorussia), a zone of relatively high W6 values appears in the Pripyat Trough. However, practically the entire Belorussian massif and the Pripyat Shaft belong to the zone of anomalously low TP and, respectively, W.

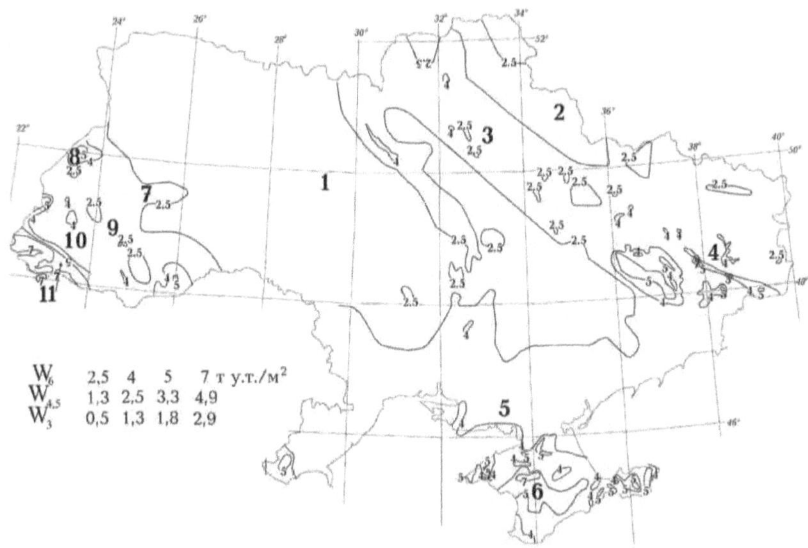

Fig. 4.10. Regional distribution of W on the territory of Ukraine.

Figures - tectonic regions of Ukraine (see Table 4.1).

The poorly studied territory of the Ukrainian Shield and the appearance of rather intense TP anomalies in the most studied parts of the Ukrainian Shield indicate the possibility of detecting here in the future at least separate zones with promising geothermal resources. Their detection is likely within the limits of the not yet fully traced Dnieper anomaly on the northeastern slope of the shield. Only the areas of the northwestern part of the Shield and the adjacent territory of the Volyno-Podolsk plate, as well as some areas in the northern part of the Shield, where it is difficult to expect the detection of even average values of heat flow for the region, at which (due to the high thermal conductivity of crystalline rocks) the level of promising resources is not reached (due to the high thermal conductivity of crystalline rocks).

A summarized scheme of geoenergy resources is presented in Fig. 4.10. It is low-detailed but illustrates the possibilities of sequential utilization of the Earth's heat from different depth intervals.

In the three basins and the central part of Ukraine with low thermal potential, the total amount of resources is about 1 trillion tons of fuel equivalent. Let us compare the obtained data with the information on the reserves of combustible minerals of Ukraine presented in Table 4.3 according to [23].

Table 4.3.

Type of fuel	Stocks	Reserves in t.c.e.
Hard coal	4.31141.10 T^{10}	3,880.10^{10}
Lignite	0.25848.10 T^{10}	0,127.10^{10}
Peat	0.0659379.10 T^{10}	0,025.10^{10}
Oil	0.01467.10 T^{10}	0,022.10^{10}
Gas	129.10 M^{103}	0,161.10^{10}
Condensate	0.00807.10 T^{10}	0,012.10^{10}
Total		**0.04 trillion tons of fuel equivalent.**

The amount of W6 exceeds the reserves of combustible fossils (mainly - hard coal) by 25 times. Taking into account the environmental advantages of geothermal energy, this allows us to assess the use of the Earth's heat in Ukraine as a very promising direction.

Evaluation of geothermal resources that can be used without reheating, producing steam suitable for power generation (fluid T - 210^0 C, discharge - 70^0 C, i.e. the discharged fluid can be used for heat supply), showed that at a drilling depth of 4.5 km, the minimum resources appear in the zone of maximum TP of the Transcarpathian sag (GTP >120 mW/m^2). At the drilling depth of 6 km, the values corresponding to the W6 values for the variant considered above (coolant T - 60^0 C, discharge - 20^0 C) are obtained: 6 - 0, 7 - 2.5, 8 - 3.8, 9 - 5, 10 - 8 t c.t./m^2 . Thus, conditions for steam production are present only in the Transcarpathian trough and in very limited areas of Crimea and Donbass.

There are very promising developments for the use of hot water (if there are already drilled wells during exploration) for steam generation when it is reheated by combustion of associated gas from pilot wells of DDW fields, methane from closed coal mines in Donbass, shale gas, etc.

It can be argued that the assumption made at the beginning of the chapter about the prospective use of geothermal energy has been confirmed by the studies conducted.

Conclusion

This brief review of the results of geothermal studies of the territory of Ukraine intentionally does not include discussion of the most complex issues of interpretation of background and anomalous parameters of the thermal field and applications of data on the heat flow and gradient to the solution of problems of prospecting for mineral deposits, studying seismicity, geodynamics, and so on. They are considered in other works of the authors [8, 11, 12, 16, etc.]. Here we have tried to give an idea of the studied thermal field, the distribution of deep temperatures and one application - the assessment of geoenergetic resources. The peculiarity of the work seems to be drawing the reader's attention to the accuracy of parameter determination (which usually remains in the shade) and the need to make corrections to the heat flux values (which is practically absent in other works).

The use of independent data to control the constructed thermal models also seems to be an advantage.

The work done allows us to confidently proceed to the assessment of geoenergy resources of Ukraine. In the extreme northeast of the territory under consideration, the map also covers a part of the Voronezh massif proper with minimal (up to the first hundred meters) sediment thickness.

e and only for one technology of heat extraction, the accumulated experience in studying the thermal field and information base allows to easily move to the study of specific fields and the use of other technologies. There is no doubt that the use of deep heat on the territory of Ukraine in the future will be more intensive than at present.

Literature

1. Alexandrov A.L., Gordienko V.V., Derevskaya E.I. et al. Depth structure, evolution of fluid-magmatic systems and prospects of endogenous gold-bearing in the south-eastern part of the Ukrainian Donbass. Kiev: IFI UNA. 1996. 74c.

2. Babaev V.V., Budymka V.F., Sergeeva T.A. et al. Thermophysical Properties of Rocks. Moscow: Nedra. 1987. 157c.

3. Buryanov V.B., Gordienko V.V., Zavgorodnyaya O.V. et al. Geophysical model of the tectonosphere of Ukraine. Kiev: Nauk. dumka. 1985. 212c.

4. Buryanov V.B., Gordienko V.V., Zavgorodnyaya O.V. et al. Geophysical Model of the Tectonosphere of Europe. Kiev: Nauk. dumka. 1987. 184c.

5. O.V. Veselov, V.V. Gordienko, V.V. Kudelkin. Thermodynamic conditions of gas hydrate formation in the Sea of Okhotsk. Geology and Mineral Resources of the World Ocean. 2006. 3. C.7681.

6. Deep xenoliths and the upper mantle. Ed. V.S. Sobolev. Novosibirsk: Nauka. 1975. 272c.

7. Gordienko V.V. Thermal anomalies of geosynclines Kiev: Nauk. dumka. 1975. 142c.

8. Gordienko V.V. Deep processes in the Earth's tectonosphere. Kiev: IGF NASU. 1998. 85c.

9. Gordienko V.V.. Density models of the tectonosphere of the territory of Ukraine. Kiev: 1ntelect. 1999. 101c.

10. Gordienko V.V. Physical properties of rocks of deep depressions. Geophys. zhurnal. 2000. 2. C.19-26.

11. Gordienko V.V. Processes in the Earth's Tectonosphere (Advection-Polymorphic Hypothesis). Saarbrücken: LAP. 2012. 256c.

12. Gordienko V.V. Modern activation and hydrocarbon deposits (on the

example of the Dnieper-Donets depression). Deep Oil. 2013. 12. C. 1688-1710.

13. Gordienko V.V. About PT-conditions in magmatic centers of the Earth's mantle. Geophys. zhurnal. 2014. 6. C.28-57.

14. Gordienko V.V., Badanov V.A., Veselov O.V., Zavgorodnyaya O.V., Shvartsman Y.G. On the accuracy of temperature measurement in boreholes. Geoph. zhurnal. 1992. 3. C.35-41.

15. Gordienko V.V., Gordienko I.V., Zavgorodnyaya O.V. Thermal field of the central part of the Ukrainian Shield. Part 1. Geophysical Journal. 1996. 1. C.52-61.

16. Gordienko V.V., Gordienko I.V., Zavgorodnyaya O.V. et al. Thermal field of the territory of Ukraine. Kiev: Znanie Ukrainy. 2002. 170c.

17. Gordienko V. V., Zavgorodnyaya O.V. Measurement of the Earth's heat flux at the surface. Kiev: Nauk. dumka. 1980. 104 c.

18. Gordienko V. V., Zavgorodnyaya O. V. Thermal field of the northeastern part of the territory of the Ukrainian SSR. Geophys. zhurnal. 1983. 3. C.27-32.

19. Gordienko V.V., Zavgorodnyaya O.V. Determination of heat flow in different geothermal situations. Standardization of geothermal studies in tectonically active areas. Makhachkala: Dag. fil. ACADEMY OF SCIENCES OF THE USSR. 1987. C.27-35.

20. Gordienko V.V., Zavgorodnyaya O.V. Thermal field of the southeastern part of the Ukrainian Shield and its slopes. Geophys. zhurnal. 1989. 2. C.39.

21. Gordienko V.V., Tarasov V.N. Modern activation and helium isotopy of the territory of Ukraine. K.: Znanie. 2001. 102c.

22. Gordienko I.V. Interpretation of Kirovograd heat flow anomaly. Geophys. zhurnal. 2000. 3. C.82-89.

23. State Geological Survey of Ukraine. Dovshnik. Kiev: GeoYform. 1999.

87c.

24. Dyadkin Y.D. Fundamentals of geothermal technology. Leningrad: LGI. 1985. 176c.

25. Dyadkin Y.D., Boguslavsky E.I., Vainblat A.B. et al. Geothermal Resources of the USSR. Geothermal models of geologic structures. C. St. Petersburg: VSEGEI. 1991. C.168-176.

26. Zabarny G.N., Shurchkov A.V., Zadorozhnaya A.A. Resources and thermal potential of prospective for industrial development of thermal water deposits of Transcarpathian region K.: ITT NASU. 1997. 150c.

27. Kadik A.A., Lukanin O.A., Portnyagin A.L. Magma formation during upward movement of mantle matter: temperature regime and composition of melts formed during adiabatic decompression of ultrabasite mantle. Geochemistry. 1990. 9. C.1263-1276.

28. Map of discontinuities and main lineament zones of the south-western USSR. M-b 1 : 1 000 000. Editor. N.A. Krylov. Moscow: Ministry of Geology of the USSR. 1988.

29. Kutas R.I., Gordienko V.V. Thermal field of Ukraine. Kiev: Nauk. dumka. 1971. 141c.

30. Lebedev T.S., Korchin V.A., Savenko B.Ya. et al Physical properties of mineral matter in thermodynamic conditions of lithosphere. Kiev: Nauk. dumka. 1986. 199c.

31. Lebedev T.S., Korchin V.A., Savenko B.Ya. et al. Petrophysical studies at high PT-parameters and their geophysical applications Kiev: Nauk. dumka. 1988. 248c.

32. Lebedev T.S., Shapoval V.I. RT-research of physical properties of rocks of the upper part of the section of the Krivoy Rog superdeep well. 3. Thermal parameters. Geophys. journal 1993. 1. C.46-55.

33. Lukin A.E. Lithogeodynamic factors of oil and gas accumulation in avlacogenic basins. Kiev: Nauk. dumka. 1997. 224c.

34. Lut R.T., Mishchenko A.V. Renewable energy sources in Ukraine - resources and characteristics. Renewable Energy. 2001. 3. C.5-8.

35. Moiseenko U.I., Smyslov A.A. Temperature of the Earth's interior. Leningrad: Nedra. 1986.178c.

36. National Atlas of Ukraine. K.: Kartografiya. 2007. 440c.

37. Physical properties of rocks and minerals. Handbook of geophysicist. Ed. N. B. Dortman. Moscow: Nedra. 1984. 456 c.

38. Shpak A.A., Efremochkin N.V., Borevsky L.V. Prospecting, exploration and evaluation of forecast resources and operational reserves of thermal energy waters. Moscow: Nedra. 1989. 89c.

39. Yakovlev B. A. Solution of problems of petroleum geology by methods of geothermia Moscow: Nedra. 1979. 142 c.

40. Armstead H. and Tester J. Heat Mining. London: E.F.Spon. 1987. 320p.

41. Fujisava H., Fyjii N. Thermal diffusivity of Mg_2 SiO_2 , Fe_2 SiO_4 and NaCl at high temperatures and pressures. J.G.R.. 1968. v.73, 14. P. 47274733.

42. Gasparik T. Melting experiments on the Enstatite-Pyrope Join at 80152 kbr. J.G.R.. 1992. 97. P.1581-1588.

43. Gordienko I. Thermal field and geothermal energy in Carpathians and Transcarpathian troug. Proceedings of the institute of fundamental studies. 2001. P.122-124.

44. Gordienko I., Gordienko V. Heat flow and geothermal resourses in Ciscarpathian basin and Volyno-Podolian plate. The Earth's thermal field and related research methods. Moskow: RUPF. 2002. P.89-91.

45. Kanamori N., Fyjii N., Mizuteni H. Thermal diffusivity measurement of rockforming minerals from 300-1100 K. J G. R. 1968. v.73, 2. P.595605.

46. Kawada K. Variation of thermal conductivity of rocks. P. 1/ Bull. Earthquake Res. lust. 1964. v. 42, 4. P.631- 647.

47. Kawada K. Variation of thermal conductivity of rocks. P. 2.. Bull. Earthquake Res Inst. 1966. v. 44, 3. P.1071-1091.

48. Kohl T., Brennil R., Eugster W. Performance investigations of a deep borehole heat exchanger. The Earth's thermal field and related research methods. Moskow: RUPF. 2002. P.126-128.

49. Seipold V., Engler R. Investigation of the thermal diffusivity of jointed granodiorites under uniaxial load and hydrostatic pressure. Gerlands Beit. Geophys. 1981. 90,1 P.65-71.

50. Seipold V., Gutzeit W. Measurements on the thermal properties of rocks under extreme conditions. Phys. Earth Planet Inter. 1980. 22. P.272-276.

51. Tester, J.; Herzog, H. Economic Predictions for Heat Mining: A Review and Analysis of Hot Dry Rock (HDR) Geothermal Energy Technology. MIT-EL 90-001. 1990. 180p.

52. Yukutaka H., Shimada M. Thermal conductivity of NaCl., MgO, coesite and stishovite unto 40 kbar. Phys. Earth Planet. Inter. 1978. v.17. 3. P.183-200.

Printed by Books on Demand GmbH, Norderstedt / Germany